# Displacement, Asylum and the City

This edited volume draws attention to the interlinked yet understudied relationship between the role of cities in dealing with international displacement and forced migration and the influence of forced migration in stimulating spatial, societal, and institutional transformations in and of cities.

In 2022, almost 84 million people worldwide were forcibly displaced. More than two-thirds of them reside in urban areas. Displacement and forced migration are an urban experience and an urban story of those seeking protection. This book helps us understanding the conditions of displaced population in cities and the way cities and urban actors respond to recent migration trends. It applies an urban perspective to the analysis of migration processes, and it provides insights into the urban governance of forced migration and asylum, the production of spaces related to forced migration, and the role of the displaced population as actors of urban change. Thereby, it covers a broad spectrum of topics including migrant dispersal, welfare and social protection, urban humanitarian policymaking and governance, neighborhood development, migrant solidarity and refugee protest, and new refugee and migrant destinations. Given the increasing mobility and displacement of human populations, this book provides a relevant prerequisite for readers interested in current urban, (forced) migration and asylum trends, and the intersections of those topics.

This book will be of great value to researchers and academics of Geography, Migration, and Urban Studies. The chapters in this book were originally published as a special issue of *Urban Geography*.

**René Kreichauf** is Postdoc Researcher at Cosmopolis-Centre for Urban Research, Vrije Universiteit Brussel, Belgium. He did his PhD at Cosmopolis and in association with the Graduate School of North American Studies, Freie Universität Berlin, Germany. His research focuses on international and urban displacement, migrant detention and refugee camps, state violence, racial capitalism, urban transformation, and marginalization processes.

**Birgit Glorius** is Professor of Human Geography with focus on European Migration Research at Chemnitz University of Technology, Germany. In her research, she focuses on recent migration phenomena in Europe, notably forced migration, and their effects on social cohesion and society formation.

# Displacement, Asylum and the City

## Understanding Migration Processes through Urban Studies Approaches

*Edited by*

**René Kreichauf and Birgit Glorius**

Routledge
Taylor & Francis Group
LONDON AND NEW YORK

First published 2023
by Routledge
4 Park Square, Milton Park, Abingdon, Oxon, OX14 4RN

and by Routledge
605 Third Avenue, New York, NY 10158

*Routledge is an imprint of the Taylor & Francis Group, an informa business*



*British Library Cataloguing-in-Publication Data*
A catalogue record for this book is available from the British Library

ISBN13: 978-1-032-46353-7 (hbk)
ISBN13: 978-1-032-46354-4 (pbk)
ISBN13: 978-1-003-38129-7 (ebk)

DOI: 10.4324/9781003381297

Typeset in Minion Pro
by codeMantra

**Publisher's Note**
The publisher accepts responsibility for any inconsistencies that may have arisen during the conversion of this book from journal articles to book chapters, namely the inclusion of journal terminology.

# Contents

# Citation Information

The chapters in this book were originally published in the journal *Urban Geography*, volume 42, issue 7 (2021). When citing this material, please use the original page numbering for each article, as follows:

**Introduction**
*Displacement, asylum and the city: understanding migration processes through urban studies approaches*
René Kreichauf and Birgit Glorius

**Chapter 1**
*Refugee urbanism: seeing asylum "like a city"*
Jonathan Darling
*Urban Geography*, volume 42, issue 7 (2021) pp. 894–914

**Chapter 2**
*Towards a parallel exceptional welfare system: the scaling down and out of forced migrants' reception in Italy*
Michela Semprebon
*Urban Geography*, volume 42, issue 7 (2021) pp. 915–936

**Chapter 3**
*Making urban humanitarian policy: the "neighbourhood approach" in Lebanon*
Romola Sanyal
*Urban Geography*, volume 42, issue 7 (2021) pp. 937–957

**Chapter 4**
*Refugees and the transforming landscapes of small cities in the US*
Pablo Shiladitya Bose
*Urban Geography*, volume 42, issue 7 (2021) pp. 958–978

**Chapter 5**
*Negotiating urban solidarities: multiple agencies and contested meanings in the making of solidarity cities*
René Kreichauf and Margit Mayer
*Urban Geography*, volume 42, issue 7 (2021) pp. 979–1002

# Notes on Contributors

**Jonathan Darling,** Department of Geography, Durham University, Science Laboratories, UK.

**Birgit Glorius,** Institute for European Studies and History, Chemnitz University of Technology, Germany.

**René Kreichauf,** Cosmopolis Centre for Urban Research, Vrije Universiteit Brussel, Belgium.

**Margit Mayer,** Center for Metropolitan Studies, Technische Universität Berlin, Germany.

**Romola Sanyal,** Department of Geography and Environment, LSE, United Kingdom.

**Michela Semprebon,** SSIIM Unesco Chair, IUAV University, Venice, Italy.

**Pablo Shiladitya Bose,** Department of Geography, University of Vermont, Burlington, USA.

# Introduction — Displacement, asylum and the city: understanding migration processes through urban studies approaches

René Kreichauf and Birgit Glorius

**ABSTRACT**

Forced migration is an urban phenomenon – an integral part of twenty-first century urban landscapes, as well as an "urban story" of those displaced. In this introductory article, we draw attention to the interlinked yet understudied relationship between the role of cities in dealing with the arrival of forced migrants *and* the influence of forced migration in stimulating spatial, societal, and institutional transformations in and of cities. Seeking in an interdisciplinary way to urbanize debates on forced migration, we analyze and structure current literature and forced migration trends, and link them to urban theory, while contextualizing the contributions of this special issue. This reveals not only how forced migration unfolds in urban contexts, but also how it changes the urban realm and the way we think about cities. Accordingly, we propose ideas for future research, adding to discussions about re-theorizing "the urban" and including urban aspects in forced migration studies.

## Introduction

Forced migration is part of contemporary urban reality. While questions in the context of forced migration are often reflected at the state level, it is the local level where forced migrants arrive and where specific policies are implemented (Glorius & Doomernik, 2020). Forced migrants' movements – socially, politically, culturally, and economically – affect and change cities and localities. New forms of spatial and social organization of urban spaces and a "refugee urbanism" have emerged (Darling, 2020, this issue). As the frontline players in the arrival, settlement, participation, and social mobility of forced migrants (Muggah & Abdenur, 2018), cities and urban actors are "becoming more and more active agents, drawing their own agenda, policy strategies and key questions/answers to challenges related to integration and diversity accommodation," but also regarding exclusion, deportation, and bordering (Zapata-Barrero et al., 2017, p. 242). These developments differ greatly; not only among regions and countries, but also between and even within urban areas. They have resulted in new forms of governing forced migration, refugee inhabitations, and the complex geographies of asylum, as well as in novel urban, governmental and non-governmental actor constellations, actions of displaced populations, and social movements.

While there has been a "modest urban turn" in refugee studies and related disciplines, acknowledging the urban dimension of forced migrants' arrival and settlement processes (Jansen, 2016), there is still a gap in urban studies with regard to uncovering the material and socio-spatial dimensions of international displacement, forced migration, and asylum at the urban scale (Soederberg, 2018). Urban theory has only recently been applied in order to analytically approach the interlinked relationship between forced migration and the production of urban space. This is surprising, because the study of urban and spatial aspects of "international displacement" (and forced migration) – the logics and purposes of spatial formations in urban areas as well as the way forced migrants deal with them and navigate urban space – can increase not only our knowledge of the constitution and governance of migration, but also that of the livelihood and struggles of the displaced in cities (Darling, 2017). Furthermore, it can reveal the conditions, logics, and practices of current global urban development; as well as the positions of, and challenges for, cities regarding the emergence of legislative and policy frameworks to respond to arrival and settlement. Especially at a time when urban studies and urban concepts are under stress (Rickards et al., 2016), the "complexity of social, political, legal and spatial reconfigurations" of forced migration and displacement is relevant for "geographers and urbanists in thinking about urban futures" (Sanyal, 2016).

In this special issue (hereafter SI) and this introductory article, we draw attention to the interlinked yet understudied relationship between the role of cities in dealing with the latest arrival of internationally displaced populations *and* the influence of forced migration in stimulating spatial, societal, and institutional transformations in and of cities. In an interdisciplinary way, we seek to urbanize and globalize the discussion of forced migration and displacement, showing how urban and migration studies concepts make sense of current trends in forced migration and explain not only how asylum unfolds in urban contexts, but also how it changes the urban and the way we think about cities. Within the SI's contributions, the combination of urban and migration studies concepts delivers conceptual and analytical value in understanding the conditions of refugees in cities and their negotiations with space. Thereby, this issue contributes to discussions on theorizing the urban and cities as they are confronted with, and are a part of, global phenomena and planetary transformation processes. By presenting fresh empirical material about cases in different parts of the world, this SI also reflects on recent developments, such as the effects of neoliberal urban politics, the rise of new destinations, the way dispersal policies manifest in cities, and questions of the right to the city. Through these perspectives, we hope to provide a basis for further conceptual works in both migration studies and urban theory.

To set the stage for the contributions in this issue, this introduction identifies and structures current trends, debates, and research fields regarding forced migration. We link them to theories in urban and migration studies, as well as to the articles in this SI. This overview illustrates ways in which to analytically approach the intertwined relationships between cities and forced migration, within and beyond this SI. We first look at the interfaces of (forced) migration studies and urban studies, and their research interests (Section 2). Next, we analyze three accounts that highlight recent trends, and contribute to the study and understanding of forced migration, its spatial and urban aspects, and its relationship to urban studies debates: Forced Migration Governance, Spaces of

Displacement, and Forced Migrants as City Makers (Section 3). Lastly, we illustrate the potential of the combination of both (forced) migration studies and urban studies for further research, sketching out an agenda for future work in these fields (Section 4).

## Situating (forced) migration studies and urban research

While the study of forced migration is a relatively young research field that emerged after the Second World War and took off in the 1980s, migration studies and urban studies look back on similar origins in the late nineteenth and early twentieth century, notably in Europe and North America. As we can see in Ravenstein's seminal work on the "Laws of Migration" (Ravenstein, 1885, 1889), and as noted by Greenwood and Hunt (2003, p. 3), "the phenomenon most responsible for initial interest in migration as a field of scientific study appears to have been urbanization." Similarly, urban research arose from interest in the growth of urban areas through different forms, structures, and acts of migration and the study of ethnic groups, as well as the effects of racialization and racial stratification in shaping the material conditions and social structure of the modern city, including poor housing conditions in "the ghetto" or in "slum communities," with this research most prominently brought forward by W. E. B. Du Bois (1899) and within the Chicago School decades later. It is also no coincidence that some of Georg Simmel's comprehensive oeuvre — including "*The Metropolis and Mental Life*" (1903/2012) and "*The Stranger*" (1908/2008) — impacted the development of both (sociological) urban studies and of migration research. Yet for a long time, urban studies disciplines have rarely included questions of forced migration in their research agendas, and migration has mostly been explored either with regard to some of the "usual suspects" (gateway cities, global cities, etc.) or concerning immigrant communities and histories within certain cities, often without links to the global structures of urbanization and city-making processes. Migration studies – forced migration/refugee studies[1] in particular – have been accused of being characterized by a nation-state-centric focus, framing the state as a monolithic and dispassionate institution that primarily functions to construct, define, and regulate migration and migrants (Haddad, 2002). Within this *methodological nationalism* (Wimmer & Glick Schiller, 2002), the state serves as the sole unit of analysis for the study of social processes, taking the role of the state or the lives and subjectivities of refugees as starting points (see Hyndman, 2000). As a result, the connection between the city and (forced) migration, including critical thinking about matters of (urban) space and its relation to social contexts and power structures, as well as about the spatial and urban dimensions of forced migration, asylum, and refuge, has remained under-theorized (Sanyal, 2016; Turton, 2005).

In the last two decades, however, there has been a turn in (forced) migration and urban research. In migration studies, this includes attention being paid to the globalization of migration and the changing directions and patterns of dominant migration flows, as well as the differentiation, feminization, and growing politicization of migration (Castles & Miller, 2014). Forced migration scholars have also started to focus on the larger constellations of sociopolitical, political-economic, and cultural processes and practices, with some acknowledging that the "ultimate cause of refugees is today's global capitalism itself and its geopolitical games" (Žižek, 2016, p. 164). This links the production of forced migration to general global migration trends and the intersectional dynamics of global capitalism – including capitalist urbanization, (de)colonialization, racialization, economic differences, poverty, and lack of development – that makes the displacement, dispossession, and regulation of forced migrants similar to that of other racialized and marginalized groups in precarious situations within the creation of the capitalist imperialist order and its "multicrisis" (Crawley & Skleparis, 2018; Georgi, 2019; Rajaram, 2018). These perspectives help us to understand that migration is never entirely free (or "voluntary") or entirely forced (or "involuntary"), consequently blurring the legal-ontological categories of the fluid diaspora of individuals labeled as "immigrants," "exiles," "refugees," "IDPs," "economic migrants," and others, and pointing to the constructed character of the labeling processes and their consequences (Zetter, 2015). Furthermore, traditional assimilationist theories and the understanding of the nation state as a "container model" (Giddens, 1995) have been replaced by more complex comparative, transnational, global, and local perspectives on migration and interpretations of "the state" (Levitt et al., 2003). As a consequence, there has been a move toward the study of the emerging international and local political forms and actors of migration governance, as well as the conceptualization of ethnicity/race, socio-cultural, economic contexts, and especially the local scale and urban space as relevant aspects of immigration, integration, and expulsion processes (Castles & Miller, 2014; Massey, 1999; Sassen, 2014).

This *local* or *urban turn* in migration studies not only "marks a shift away from solely examining the national level to studying the growing divergence between national and local immigration policies" (Kühn, 2018, p. 1748). It also emphasizes the role of cities and of urban space for migration (and vice versa), highlighting the connection between migration and processes of urban transformation, and emphasizing the role of migrants in the making of urban spaces and urban politics (Hall, 2015). Consequently, concepts of locality and research into the relationship between migrants and urban space in constituting and challenging the local, national, and global have been conducive to research dealing with the urban-migration nexus in the era of globalization. For example, urban and migration scholars such as Brickell and Datta (2011) highlight the importance of urban space for the migration experience, as well as the emergence of transnational/translocal geographies and communities – a comprehensive effort to conceptualize migration and urbanism in what Smith (2001) has introduced as *transnational urbanism*. Against this backdrop, scholars such as Brettell (2003) have argued for viewing the *city as a context* to understand migration processes and immigrant communities as part of the institutional structures and the path-dependent history of specific cities with regard to migrant settlement. Expanding these approaches, others – most prominently Glick Schiller and Çağlar (2011) – have pointed to the relationship between migrants and urban restructuring and rescaling processes, placing migrants as active agents in these processes. Work on urban (as opposed to national) citizenship has contributed to the understanding of migrant identities and livelihoods, and their negotiations with urban space and politics (see Bauder, 2017; Varsanyi, 2006). In addition, studies into the migration-urban nexus are also driven by conceptualizations of the local (or multilevel) governance of migration (Scholten et al., 2016), (im)mobility (Urry, 2007), and the concept of *arrival infrastructures*, seeking to merge sovereign and agency-oriented approaches in multiscalar urban settings relevant for the organization, production, and power negotiations of migrants and urban space (Meeus et al., 2019; Nettelbladt & Boano, 2019).

While we cannot neglect that displacement also takes place in non-urban areas (notably in the wake of climate-induced land degradation or of land-grabs catalyzed by neoliberal policies worldwide), the interrelations between capitalist urbanization and restructuring processes, global migration flows, and global/local governance (Harvey, 1996) have become most evident in current forced migration trends and debates. A burgeoning, yet fragmented, branch of scholarship on the urbanization of refuge and the global movement of forced migrants to and within urban areas aims to make sense of the complex urban and migration systems and structures (Rickards et al., 2016). Especially since the 2000s – and with UNHCR's acknowledgment of, and subsequent changing approaches toward, the around two-thirds of refugees who reside in urban areas – scholars have been focusing on *urban refugees* in the Global South, and the challenges they face in, and their contributions to, urban environments (e.g. Grabska, 2006; Ramadan, 2013). In the Global North, the affirmation that forced migration is in fact an urban phenomenon – and one that poses questions for cities and urban governance – has been particularly pushed by increasing immigration since the early 2010s and the so-called refugee crisis. Since then, scholars have been concerned with issues such as the local governance of, and policy-making toward, (forced) migration and asylum – including dispersal, reception and accommodation practices, or the struggle for rights, place-making and spatial appropriations of forced migrants – often seeking relevant conceptual approaches to analyze these processes (e.g. Kos et al., 2016; Lidén & Nyhlén, 2015; Meret & Diener, 2019). In addition, a growing body of work pays attention to the way different forms of forced migration also affect, urbanize, and diversify rural areas and smaller towns, which have become relevant sites for studying both the "management" of migration and the expansion of capitalist urbanization (see Bose, 2020, this issue).

## The city and forced migration: an interdisciplinary debate about policy, space, and people

Picking up on this overview, in the following sections we discuss research works and current debates in (forced) migration and urban studies that exemplify scholarly interest and dominant trends regarding the interrelations between forced migration, the city, and the urban realities of asylum. This analysis contributes to the growing scholarly ambition to "better unpack the urban character of forced migration" (Darling, 2017, p. 178) that has been prominently pushed by Romola Sanyal (2012, 2014) and Jonathan Darling (2017), – two authors in this SI, whose comprehensive work and ideas on the urbanization of refuge and the role of cities have profoundly inspired and shaped research in recent years. Continuing and expanding their and other scholars' dialogs, we investigate trends along three components – policy, space, and people – in these strongly intertwined areas of research. We exemplify the contribution of literature in these fields to the broader debates regarding the city and forced migration, and explain how the SI articles advance conceptual and empirical knowledge in these fields.

### Forced migration governance

Forced migration and asylum are broadly governed under the international protection regime of the 1951 Convention Relating to the Status of Refugees, and its 1967 protocol, as well as through the respective national asylum laws and policies. However, scholars such as Guiraudon and Lahav (2000) have noted that during the last three decades, the scale of refugee and asylum policy-making has shifted both upward and downward. First, a major process of upscaling has been manifest in regional approaches and policy frameworks, international treaties, transnational (or transregional) agreements, or the externalization of border management through contracts between receiving, transit, and sending countries. Second, attention paid to the topic of refugee migration and reception has shifted downward, bringing the local level to the fore as a relevant place for governing forced migration (Glorius & Doomernik, 2020). Particularly since the so-called refugee crisis of 2015, state restructuring processes have resulted in a transfer of policies and practices from national to municipal scales (Bauder, 2017), in which the city has become "the preferred space for the implementation of migration policies" (Manfredi-Sánchez, 2020, p. 3). Research on these processes often uses multi-level governance (MLG) or regime approaches, which conceptualize political systems as multi-faceted and multi-scalar relations, characterized by the dispersion of state authority across interdependent, and yet to a certain degree autonomous, public authorities and nonpublic organizations located at different levels of government (Hooghe & Marks, 2001; Horvath et al., 2017).

Within these debates, there is a strong focus on the structures and organization of reception, dispersal, and (re)settlement, as processes that exemplify interlocking policy procedures fundamental to asylum systems in the asylum process. These also concern the arrival, residential trajectories, and mobility of forced migrants in receiving countries in the Global North (Schuster, 2005; Wren, 2003). As Michela Semprebon (2021, p. 1) highlights in this SI, dispersal and reception measures are "part of an articulated framework of [. . .] normative arrangements," which characterize asylum regimes at international (e.g. UNHCR refugee resettlement), regional (e.g. EU emergency relocation scheme), state (e.g. national allocation systems), and local (dispersal within a county or municipality and placement in particular areas) governing scales. Thus, while dispersal is not per se an "urban" policy, it does subsequently result in coercively placing forced migrants in certain localities, thereby transferring issues of maintenance, welfare support, social assistance, and the control of forced migrants to the scale of cities and municipalities. This takes place "under the auspices of a neoliberal reorganization of public policy, by the general downloading of responsibility to lower levels of governance" and in the name of "burden sharing" (Schmidtke, 2014, p. 87).

Cities and municipalities respond differently to such top-down *governmental processes* (Darling, 2017). This is because, first, reception often depends on local resources, including the availability of housing and the structure of housing markets, municipal capabilities (and willingness), civil society structures, volunteer work, and local (political and societal) experiences with immigration (Glorius et al., 2021; Myrberg, 2017). A lack of these resources may lead to cities being challenged by arrival and settlement, placing and dispersing forced migrants in the city "somewhere" and "somehow." Second, scholars such as Zapata-Barrero et al. (2017), Werner et al. (2018), and Seethaler-Wari (2018) find that urban actors are becoming less recipients of policy decisions from above, and increasingly agents who (re)define their role in refugee politics and humanitarian action, and pursue their own agendas. This includes actively participating in allocation and dispersal processes, and creating their own strategies for governing arrival and for matters of "integration," and is reflected in the burgeoning scholarship about urban migration policies and the ways municipal governments negotiate asylum and settlement. Here, some scholars see that cities formulate "immigrant-friendly" solutions to matters of reception, housing, and participation (Neis et al., 2018; Steiner & Reinhard, 2018). In particular, concepts of, and scholarly interest in, "sanctuary cities," "cities of refuge," "welcoming cities," or "cities of safe harbor" elucidate the (potential) role and limits of urban policies and urban actors for providing (or not) means for residency, inclusion, and (social-legal) protection of forced and other migrants, as well as for the rise of transnational municipal networks in the field of migration and asylum (e.g. Kreichauf & Mayer, 2021, this issue; Kron & Lebuhn, 2020; Manfredi-Sánchez, 2020; Oomen, 2020). They may offer forms of *urban citizenship* that draw on *the right to the city* through which claims to rights, services, protection, and political participation are made (Baubock, 2003).

Others, however, question the positive implications of such policies. Critical scholars aim to "debunk the myth of sanctuary cities" (Tramonte, 2011), by revealing their nature as a means for the continuation of sovereign control, including disciplining, policing, and regulating migrants, and/or as tools for promoting "diversity" as economic exigency. In recent research, U.S. scholars have illustrated that entrepreneurial development agendas are at the core of local governments' decisions to become proactive in immigration politics (Pottie-Sherman, 2018; Watson, 2019), adding to research about neoliberal urban policy-making, "migration industries," and the influence of entrepreneurs and private actors that – motivated by the pursuit of commercial gain – participate in migration governance and the capitalization of (forced) migrants (Fernandez, 2007). Some European research also demonstrates how local authorities actively refuse the reception of forced migrants (Marchetti, 2020), or intentionally locate them in socially deprived urban areas (Phillimore & Goodson, 2006). However, as work by Kos et al. (2016) shows, there is often a simultaneity of inclusionary and exclusionary elements that characterize urban policies and the urban life of forced migrants as a result of – sometimes adversarial and at other times cooperative – negotiation processes.

For the conceptualization of cities and forced migration in this SI, this has several implications. First, it expounds that the city is a "networked and contested space of immigration policy" (Mayer, 2018, p. 232), or a *battleground of asylum* (Ambrosini, 2021), where different actors from multiple scales interact both vertically and horizontally, cooperating and/or engaging in conflict. Thus, the case of asylum and forced migration exemplifies that cities perform "contorted scalar gymnastics" (Rickards et al., 2016, p. 1537), because they are drawn into alliances and are entangled within, act on, and influence different scales of policy-making, while their actions themselves are a result and part of multiscalar negotiations. Second, we see that urban policies and multiscalar negotiations not only consist of the public political claims of major (urban) government stakeholders, but also of a large variety of private and civil society actors and resident and migrant populations with their specific attitudes, interests, and institutionalized and informal responses to newcomers. Cities then have to navigate "a complex 'tightrope-walk’,” as they share and compete for power with supra-local government and non-governmental institutions, while also having to respond to demands emerging from multiple actors from “below” (Bazurli, 2019, p. 344).

The articles in this issue further the conceptualization of the negotiations of different actors and authorities, and their collaborating and competing interests, as a relevant frame for research and action. For example, Kreichauf′s and Mayer′s (2021) contribution concerning “negotiating solidarity cities” reveals the ambiguous and complex ways in which diverse actors – including local governments, welfare associations, churches, private entrepreneurs, nonprofit organizations, volunteers, and activists, as well as refugees and other migrants themselves – and their constantly changing practices regulate and condition solidarity in the urban realm. In a similar (yet conceptually different) fashion, both Darling (2020) and Semprebon (2021) use the context of “refugee and migrant” (dispersal) and placement as political processes to analyze the complexity, political contestations, and divisions of the multiplicity of authorities involved in these processes. To do so, Darling (2020, p. 2 f.) explores the notion of “seeing like a city” – as a counter-position to seeing like a state – to establish a *refugee urbanism* approach in order to examine the conjunctures of factors, actors, and unequal power structures that “shape how refugee subjectivities intersect with, are remade by, and themselves serve to constitute and rework, the urban.” Semprebon (2021) studies dispersal through the rationales of welfare policies in Italy, revealing the development of parallel and exceptional welfare systems that unfold in the local context of cities and impact local policy-making as well as forced migrants’ trajectories. She explains that these developments are a result of, and further contribute to, the restructuring of the welfare state and the downscaling of welfare policy. Sanyal (2021) analyzes the neighborhood approach in Lebanon as an example of the way humanitarian organizations respond to the settlement of forced migrants in urban areas, illustrating that despite being a global humanitarian policy, the neighborhood approach exemplifies how it is developed through local experiments and negotiations of NGOs and local actors. Lastly, Bose’s (2020) analysis of the reorganization of the U.S. resettlement program demonstrates the changing settlement patterns of refugees who are increasingly placed in and are moving to rural areas and smaller towns. He reveals the rationales for this shift – the perception that racialized minorities and forced migrants are enablers of urban renewal processes – as well as the strategies and challenges that are linked to arrival in new destinations.

The articles in this issue exemplify multiple claims to urban policy-making, authority, and power, and how they are produced in different collaborations, competitions, and contexts in the urban asylum arena. Furthermore, they elucidate how these negotiations – especially in regard to movement, rights, and mobility – influence and result in various and changing spatialities and geographies. This illustrates that forced migration is (re-) constituted through spatial productions of power (Sigona, 2015), an aspect that is significant with regard to understanding spatial formations related to forced migration that we discuss in the following section.

### ***Spaces of displacement***

There is one area in which forced migration studies and urban theory have been intervening for a while: the *refugee camp* and its analogies. The study of these spaces has occupied many urban and forced migration scholars alike, who often describe the refugee camp and its variations as a form of the spatial ordering of mobility and migrations, and "a standardized, generalizable technology of power in the management of mass displacement" (Malkki, 1995, p. 498). However, the literature in this field reveals some theoretical slippages and limitations. For example, conceptualizations and notions of the refugee camp appear to have a Global North and South geography. In the South, "refugee camp" often describes the spontaneous development of a temporary settlement and "a kind of 'humanitarian space', an attempt by the international community to institutionalise a state of protection and relief for refugees in an enduring but ultimately temporary way" (Ramadan, 2013, p. 69). In the North, scholars frequently refrain from using "camp," often in favor of normative and administrative terms (asylum center, asylum accommodation, reception center, etc.) to define these spaces as government-planned and organized places of care, custody, mobility control, and state exclusion (or "integration"), and to illustrate their institutional nature as part of national migration and asylum laws, dispersal regimes, and asylum procedures (Szczepanikova, 2013; Van Liempt & Miellet, 2021). By contrast, "camps" in the North are often considered to be informal encampments, makeshift camps, and tent cities (such as the *Calais jungle*), developed by transiting migrants and as a response to restrictions to their movement, mobility, and rights, even though they share "a complementary, almost symbiotic relationship" with institutional state encampments (Martin et al., 2020, p. 743).

On the one hand, these distinctions illustrate that the geographies and socio-spatial formations of refugee camps vary regionally, but also nationally and locally – especially with regard to living conditions, functions (emergency help, detention, reception, accommodation, etc.), levels of exclusion, sovereignty, and institutional responsibility. Thus, they represent highly diverse settings that govern forced migrants to different degrees (Turner, 2016). On the other hand, they reveal the difficulty in establishing a global language; including a focus on some conceptual similarities, especially concerning the positioning of camp spaces at the extremes of autonomy and heteronomy, "formality and informality, mobility and immobility, permanence and impermanence" (Grbac, 2013, p. 3; see also: Malkki, 2002; Martin et al., 2020). To explain these variations, scholars have for a long time turned to Agambenian ideas of the camp as a *space of exception*, immobilization, and deprivation of rights: the practice of sovereign power to reduce forced migrants to "bare life." This is obviously short-sighted, given the settlement history and social complexity of camps in the Global South such as Cox's Bazar in Bangladesh, Zaatari in Jordan, or the Dadaab Complex in Kenya. These can perhaps be better understood as small towns in their own right, stabilized by local governance structures and the individual agency of their forced inhabitants. Consequently, this perspective has been contested by several researchers who stress the complexity of camp spatialities, describing them as sites in which new identities, acts of agency, political action, and resistance are formed and practiced (Malkki, 1995; Martin, 2015; Ramadan, 2013; Sanyal, 2014), rather than characterizing camp inhabitants as passive victims. Sigona (2015, p. 1 f.), for example, shifts "the analysis of camps away from exceptionality" toward the more agent-oriented approach of *campzenship* to capture "the specific and situated political membership produced in and by the camp." Similarly, scholars investigating camp-like asylum accommodation (Dempsey, 2021; Ghorashi et al., 2018; Hartmann, 2017) or migrant detention (Campesi, 2015; Kreichauf, 2021a) in the Global North have questioned the unfettered power of the state and the biopolitical control of forced migrants by making the importance of spatial appropriation practices, lived experiences, social relations, agencies, and the *actual* power relations and negotiations in camps between a variety of actors the subjects of research (often by applying Lefebvre's or Bourdieu's critical spatial theories). In that way, the camp is viewed as a relational and socially produced space, "where rights, entitlements and obligations are reshaped, bended, adjusted, neglected and activated by and through everyday interactions" (Sigona, 2015, p. 11).

These understandings open the door for conceptualizing and juxtaposing the camp within broader urban studies frameworks and theories of space. For example, scholars such as Agier et al. (2002), Ramadan (2013), and Sanyal (2012) see a topological relationship and draw direct connections between the camp space and urban space, city development, urban economies, livelihood, or community building, highlighting the quasi-urban nature of camp environments as well as the somewhat organic urbanization of camps. Refugee camps can become *potential cities* or *part of cities* (Katz et al., 2018); they spill over into or become cities and gain permanence and prominence for urban dwellers (Fawaz, 2016a). In the light of this work, Martin et al. (2020, p. 755) put forward that "the urban has therefore become an important analytical framework to interrogate the refugee camp and its spatialities." There is growing interest in studying the embedment of camps in *border spectacles* (De Genova, 2013), and concerning the rescaling, diversification, and internalization of borders and border-making practices in cities (Lebuhn, 2013). In this regard, scholars such as Fontanari (2015) and Hamann and El-Kayed (2018) emphasize that camps are connected to both the making and unmaking of global, national, and local borders and the formation of *urban borderscapes* (Bastide, 2014) with distinct spatial, legal, and temporal features within and between cities. In these processes, the city itself becomes a site of bordering (Darling, 2017), affecting the way forced migrants are mobile, negotiate urban life, and "experience both the camp and the urban environment around them" (Picker & Pasquetti, 2015, p. 685). This illustrates that the logics of the camp (confinement, control, policing, etc.) and nation states' enforcement measures, as well as the struggles over agency, power, movement, and claims to political and social life in cities, are related to urban border-making processes, and that the border has crawled into "every pore of the urban fabric" (Lebuhn, 2013, p. 43).

These developments "blur the boundaries between camps and cities" (Sanyal, 2017, p. 124), and scholars (including the authors of this SI) increasingly apply an *urban lens* to understand how spaces related to forced migration are produced, negotiated, and contested in and through cities. Casati (2018), Nettelbladt and Boano (2019), and Porter et al. (2019), for example, show that camps are part of wider urban processes and increasingly incorporated into broader urban planning and urban governance approaches. This takes us empirically and conceptually beyond the camp as the most prominent site for studying the politics, realities, and spaces of asylum and displacement, toward expanding and overlapping spatial realities in the era of intensified global displacements (Huq & Miraftab, 2020). Because, as Sanyal (2014, p. 558) argues, "refugee spaces are emerging as quintessential geographies of the modern," scholars consequently explore other emerging forms of non-camp spaces, refugee inhabitations, and spatial productions related to forced migration, challenging dominant perceptions of spaces of displacement and refuge. Driven by studies in the Global South, researchers are interested in self-settled spaces, informal settlements and their variations, social practices, and mobility of forced migrants in and across urban spaces and *en route* (Fawaz, 2016b; Ikizoglu Erensu & Kaşli, 2016), as well as their appropriation and invention of, and contribution to, the production and transformation of urban spaces for political, social, and economic participation (Grabska, 2006; Huq & Miraftab, 2020; Lyytinen, 2015).

The articles in this SI analyze spatial productions related to forced migration from a variety of scalar perspectives and disciplines. Bose's (2020) study explains how small towns have emerged as new spaces for the displaced, where the spatial productions of and for refugees are embedded in local logics, narratives on local urban development, and policy approaches. Sanyal (2021) contributes to this discussion by changing the focus from the camp to the neighborhood level, studying the application and limits of a neighborhood approach – a whole-of-community design, which is well known in the context of urban and social planning – to the topic of refugee relief. Her article elucidates the changing conceptualizations of urban territories for refugee care, and the increasing impact of humanitarians in defining and making these territories. In his article, Darling (2020) expands the meaning of spaces of displacement, as his case studies highlight experiences of forced mobility and immobility, and their effects for migrants. He argues that dispersal and accommodation can be interpreted as hegemonic practices, in which forced migrants are "managed, monitored and controlled according to conditions defined by the sovereign authority" – characteristics similar to the refugee camp. Semprebon (2021) illustrates some general trends about institutional refugee camps in Europe: the (re-)organization of reception and accommodation has experienced a reduction of funding and lowering of minimum standards, and it has been largely downscaled to regional and municipal authorities that contract private and third-sector actors. This not only results in limited accountability for the living conditions of forced migrants in these camps, but also in protracted displacement and permanent temporariness, because of, poverty, discrimination, racism, barriers to finding housing, and dependence on local authorities' fragile welfare services. In contrast to the study of institutional spaces, Kreichauf and Mayer (2021) explore the way social movements and forced migrants produce urban space – e.g. in the form of protest camps, squatting, and housing cooperatives – as articulations of rights, identities, and collectivities.

This leads us to the next section, where we focus on forced migrant subjectivities, migrant activism, and urban social movements that are often the consequence of or that result in specific urban practices and spatial productions.

## *Forced migrants as city makers*

Migration and forced migrants impact on urban development and vice versa. Forced and other migrants have always been actors of urban change – from "building 'ethnic' precincts and neighborhoods, contributing to evolving and hybrid urban cultures, triggering gentrification and urban renewal, making various claims to recognition as urban citizens and even becoming part of the 'branding' of the cosmopolitan or global city" (Robertson, 2014, p. 230). As Harvey (2000, p. 25) argues, "the massive forced and unforced migration of people now taking place in the world [. . .] will have as much if not greater significance in shaping urbanization in the 21$^{st}$ century as the powerful dynamic of unrestrained capital mobility and accumulation."

Current research shows in different ways that forced migrants, while they are exceedingly vulnerable, develop agency and practices of resistance. Even though the state aims to manage and govern migration to deploy power over forced migrants while putting them in precarious subjectivities and spaces, these positions and spatial productions can also be starting points for "the weak [to] create their own places within those places; making them temporarily their own as they occupy and move through them" (Jewkes, 2013, p. 128). Thus, forced migration is conditioned by forced migrants' socio-spatial engagement in the urban context and their – often unexpected – agency to appropriate and resist institutional structures and connected subjectification (Ghorashi et al., 2018; Ikizoglu Erensu & Kaşli, 2016; Landry, 2015). This exemplifies that forced and other forms of migration first and foremost follow logics, motivation, and trajectories of their own, to which states and state actors often respond with regulatory spatial measures (Papadopoulos & Tsianos, 2013). Forced migration is therefore a process that includes a constituent force in the formation of polity and social life – an understanding that has been advanced in the literature about the *autonomy of migration* (Moulier Boutang, 2007), or – more recently – the *autonomy of asylum* (Tazzioli et al., 2018). In this context, researchers argue that forced migrants' tactics not only contest the regulatory orders that have been developed in reaction to their movement, but that they can "delineate notions of how the city is to be inhabited and used" (Simone, 2004, p. 407), and demand participation, presence, and mobility that may be reinforced through the negotiations of urban life.

These processes are manifold and reveal the making of the urban: First, the arrival, presence, and mobility of forced migrants can transform and recondition urban debates, political strategies, and local governance. In different European, North American, and Latin American cities, the increased arrival and movement of forced migrants in recent years has strongly influenced and rescaled urban policy-making, where issues of solidarity, participation, and affordable housing, as well as the strengthening and access to municipal services for all urban residents, have gained momentum (Agustin & Jørgensen, 2019). Such developments demonstrate that forced migrants' presence, the articulation of political subjectivity, and claims to have rights can even reposition "the immigrant and the citizen as urban subjects, rather than essentially different subjects" (Sassen, 2013, p. 69). Second, as Kreichauf and Mayer (2021) show in their contribution to this SI, forced migrants are political agents who actively participate in the urban governance and repositioning of cities through protest, activism, and place-making processes. Forced migrants establish political subjectivity that is often an outcome of, or result in, mobilization, protests, and activism. In relation to this, forced migrants further quite literally "make" urban space, transforming the locations and material spaces they occupy, and the functions, purposes, and meanings of these places. This can constantly change the mobility and position of the people that inhabit and use the spaces, producing a multidirectionality of newcomers' trajectories (Meeus et al., 2019).

The contemporary migrant metropolis provides the context, platform, and socio-spatial and political opportunities in which claims are made possible (Meret & Diener, 2019). Through forced migrants' acts of protest and resistance, they claim and make space. Political space is thus opened up, through which these politicized groups act and make use of different spatialities to further their cause and constitute themselves as political subjects, often by connecting within the overall urban environment and by interacting and networking with local residents or other marginalized groups in the city (Belloni, 2016; Ikizoglu Erensu & Kaşli, 2016; Landry, 2015). These spatialities can transform over time, constantly contributing to and reconfiguring urban policy debates and urban space. Thus, research in this field shows that spatial productions related to forced migrants represent political spaces that simultaneously prompt institutional logics and questions of immigrant integration. They also represent symbolic, cultural, and economic connectivity between, and embeddedness within, geographies that are specifically local, yet spatially global (Brickell & Datta, 2011). Overall, forced migrants' agency, their political subjectivities, and their city making processes reveal broader questions about the relationships between cities and citizenship and the geopolitical role of urban space (Landry, 2015).

In this SI, the authors discuss the role of forced migrants as city makers in inspiring ways. Sanyal (2021) delivers a very specific view on how forced migrants can appear as urban agents, showing that their role as city makers is conditional on various stakeholders and power relations. She disentangles planning and policy-making cultures, cultures of cohabitation, and underlying national political conflicts, and raises relevant questions regarding the representation of vulnerable groups in policy and planning processes, and whether those processes – participatory and deliberative – could enable socially just outcomes. Bose (2020) explains that the arrival of forced migrants in small, and often struggling, towns is frequently met with expectations for urban revitalization in a context of strong demographic and economic decline. Local authorities are motivated by economic growth agendas and promote inclusionary values, an overtly positive imagination of diversity, and welcoming strategies toward immigrants, hoping that resettled refugees may contribute to the urban regeneration and rebranding of these towns. His analysis points to the connections between migration politics and city making, exemplifying the increasing critical research potential with regard to the immigration-revitalization nexus (Pottie-Sherman, 2018; Watson, 2019), but also to the specifics of the urban in different types of cities and regions – such as small towns and mountain areas – that all have particular connotations in economic and historical terms (see also Glorius et al., 2021; Kühn, 2018). Kreichauf and Mayer (2021) reveal that forced migrants are political agents who actively participate in the urban governance and the repositioning of cities through protest, activism, and place-making processes. Analyzing the practices of forced migrants (among other actors and trends), they demonstrate how their mobilization implies a political strategy of creating visibility, thereby challenging authorities and influencing policy agendas. However, they also demonstrate that local authorities' counterstrategies of policing, controlling, and bureaucratizing in reaction to the visibility and actions of forced migrants can become an existential threat. Similarly, Darling (2020) shows that the "minor politics" of forced migrants (e.g. monitoring and recording inadequate and unsafe housing conditions, and anti-deportation activism in support of fellow forced migrants) and their search for recognition of political presence are often the starting point for building opposition to government policy, and for creating alliances of interest between activists, local authorities, and forced migrants that impact on wider urban policy and development processes.

## Toward a forced migration agenda in urban studies

This SI demonstrates that combining forced migration studies' focus on the human experiences of displacement and dispossession over space and time (Fiddian-Qasmiyeh et al., 2014) and urban studies' aim to capture "geographically variegated, complex, hierarchical, multi-scalar and always processual socio-spatial assemblages and territorial configurations" of urbanization (Rickards et al., 2016, p. 1531) can offer new possibilities for theory and application. The articles in this issue advance current debates and enable us to understand both forced migration and urban processes through explorations and imaginations of the urban and the city. In the final section, we spell out some implications of the SI for future research that may help to further establish forced migration in urban analysis and offer an urban lens for forced migration studies.

### *Studying post-arrival geographies, (im)mobility, and temporality*

Forced migration research, particularly work that has emerged since the so-called refugee crisis in the Global North, has paid significant attention to reception and resettlement processes. Yet we know little about the factors that shape the trajectories of forced migrants once they are granted residency and a form of state protection, including their struggles to find affordable and permanent housing and decent work. There is also little knowledge about forced migrants' experiences when they are denied asylum, become "illegal," or are deported, and have to navigate "illegality" or a life after deportation to their so-called "home" or "third" countries. Therefore, future research should look further into what El Moussawi and Schuermans (2021) term *post-arrival geographies*, including micro-scale movements, while paying attention to the particular circumstances, political, economic and social structures, and challenges the diverse groups of forced migrants face when, for example, accessing housing markets, urban infrastructures, work, and local economies. This may also be relevant for understanding the livelihood and movement patterns in new migrant destinations, continuing Bose's (2020) work in this issue, or for looking into the way changes or failures of local policies – such as the neighborhood approach in Lebanon that Sanyal (2021) analyzes – impact mobility, or the reframing/reimagining of neighborhoods and of the urban body as such.

In particular, studies that apply biographical or migration trajectory approaches could be beneficial in order to systematically examine the effects of individual characteristics and family composition as well as for framing structures of the mobility of forced migrants over time. This may not only encompass studies on the ways forced migrants negotiate (im)mobility while deciding whether, when, where, and how to be mobile or to be fixed to places in relation to territorially based logics of the state and NGOs (Weidinger & Kordel, 2020). It could also concern the construction of transnational and translocal spaces through the multiple movements of forced migrants, as well as the process of illegalization and deportation and the livelihood of illegalized and deported migrants and their continued ties to urban networks. Lastly, studying post-arrival geographies opens up the possibility to examine the longer-term progression and transformation of urban (arrival) spaces and infrastructures themselves, and how these intersect with historical forms of migration, long-term settlement, racism, and displacement that are inherently place based and place specific. Such undertakings may incorporate comparative research about the role of different urbanities and localities (urban areas, rural areas, declining cities, prosperous cities, etc.), as well as the specific conditions and actors involved in influencing forced migrants' trajectories (De Hoon et al., 2021), helping us also to learn more about why urban responses differ so markedly in various contexts.

### *Re-theorizing displacement and urban marginalization*

Forced migrants' post-arrival trajectories are often linked to continued displacement and marginalization processes. This is because experiences of coercion, processes of dispossession, and *un-homing* (Elliott-Cooper et al., 2020) do not cease on entering a receiving country and city. Instead, forced migrants may experience *circuitous displacement* (Bhagat, 2020), not only on "arrival," but often long afterward. This ongoing displacement is due to two reasons. First, the described spatial technologies and the multiplication of urban borders prolong and expand displacement in the city of arrival (Hamann & El-Kayed, 2018; Kandylis, 2019; Kreichauf, 2021b; Semprebon, 2021, this issue): they forcibly displace the already displaced; often in instutionalized spaces (camps, detention) or "left by themselves" in spaces of organized abandonment. Second, forced migrants face housing shortages, wider urban patterns of exclusion, and raced markets "with the material ramifications of refugee management within disciplinary logics of market-oriented governance" (Bhagat, 2020, p. 2). They are dramatically affected by the transformation of housing markets in urban centers, and they have "to traverse urban spaces marked by growing levels of internal displacement (evictions and homelessness)" (Soederberg, 2018, p. 924). Because there are historical and imperial dimensions that connect global cities and racialized forms of displacement (Danewid, 2020), continued displacement takes place in relation to capitalist value regimes, the position of forced migrants within these regimes, and the wider history of racial exclusion and marginalization in cities.

This has two important implications for future urban and (forced) migration research. First, we see the need to advance an interdisciplinary theory of displacement. The reason for this is that even though displacement strongly relates to urbanization processes, which are frequently the starting point for dispossession and global forced migrations (Van Vliet et al., 2017), both urban studies and forced migration research usually work with separately developed frameworks, scales, and forms of displacement. In urban studies, displacement refers to the commodification and financialization of land and housing, where urban redevelopment projects spawn displacement processes of disadvantaged, and often racialized urban residents (see Marcuse, 2013; Yiftachel, 2020). Migration studies seem to be concerned with broader processes of coercion and expulsion on the scale of regions or countries, triggered by conflicts, wars, persecution, or development – processes that result in the forced (and often transnational) movement of people who experience the loss of home and social networks (Hyndman, 2000; Koser & Martin, 2011). An advanced and interdisciplinary theory of displacement could focus on the relationship between displacement and the displaced, and urban processes, within urban areas and in relation to the global urban world, while combining the rich theoretical ideas of displacement as a process that "ruptures the connection between people and place" from both urban studies and migration research perspectives (Elliott-Cooper et al., 2020, p. 492).

Second, because forced migrants "are often studied as though they have no relation to the racial and class structures of the societies in which they reside" (Rajaram, 2018, p. 627), this approach enables us to draw connections to marginalized urban populations and the "urban poor," "slum dwellers," and the urban outcasts who "face a series of deprivations that can be productively compared against the condition of refugees" (Sanyal, 2012, p. 639). As part of "racialised and classed minorities," forced migrants are equally vulnerable to the spatial effects of gentrification, "including 'dumping', 'containment' and 'stigmatisation'" (Hill et al., 2021, p. 260), and therefore refugee spaces can be considered as spaces of urban marginality. In this regard, Huq and Miraftab (2020, p. 353) conclude that "contemporary cities are sites that receive displaced populations, as well as engender displacement through urban development, on scales of intensity that are comparable to international displacements." Their argument raises the questions of how, to what extent, and under which conditions (and possible limitations) we can equate and analytically compare different forms and practices of displacement and marginalization, as well as racial urban policies and structures.

### Rethinking forced migration and the city

After bringing forth several open research issues in the context of forced migration, we need to ask how we can embed the indicated research agenda into recent promising meta-approaches of urban and (forced) migration studies. Both disciplines have developed a fruitful and critical debate over the last decade; a debate that touches disciplinary foundations and leads toward a critical reappraisal of central categories and claims.

In the context of (forced) migration studies, new inspirations have entered via post-colonial/de-colonial and post-migration approaches, which apply a counterhegemonic perspective to knowledge production. While the post-colonial approach constitutes a rupture in historiographic meta-narratives and turns to the counter-perspective of the colonized, the post-migration approach questions the historical exceptionality of migration and thus opens up the perspective beyond the binary categories of "migrant/non-migrant" toward underlying separating categories such as "culture," ethnicity, religion, or social status (Hess, 2015; Römhild, 2015). By mirroring the racist and excluding practices of the "population majority," the post-migration approach aims to challenge dichotomous perspectives and invites a re-interpretation of migration phenomena and their societal consequences (Yildiz, 2014). Based on those critical approaches, research into forced migration could focus more systematically on the effects of categorization processes instead of selecting the research subjects solely based on their protection status, thus essentializing normative categories (Hamlin, 2014; Müller-Funk, 2020). Apart from questions connected to forced migrants' legal status and the fact and effects of their coerced movement, we will probably – turning to the urban space and to homemaking processes – find many commonalities with other marginalized groups, such as the spatial effects of social and political exclusion as well as empowerment and resistance (e.g. Belloni, 2016), as is also shown in the contributions to this SI (e.g. Darling, 2020; Kreichauf & Mayer, 2021). This would further facilitate focusing on relevant differences based on categories such as class, gender, age, sex, nationality, or social capital (e.g. Koçak, 2020; Suerbaum, 2020; Yurdakul & Korteweg, 2021). In addition, a closer look at the spatio-temporal contingency of forced migrant movements and transnational/translocal networks would enable a better understanding of the connections between phenomena of forced mobility across spatial and temporal orders. For this purpose, ethnographic studies following migrants on their trajectories would be helpful (e.g. El Moussawi & Schuermans, 2021), as would studies that examine the historical development of local (forced) migrant communities, arrival infrastructures, and their spatio-temporal transnational networks.

In urban studies, we could build on the *methodological urbanism* critique, which pleads for a relational understanding of urban processes. One strongly debated methodological innovation is Brenner's (2018) concept of *planetary urbanism*, which aims to develop a generic concept of urban processes and urbanity. This concept and its critical appraisal suggest a radical deconstruction of pre-formulated categories, such as the scale of the "urban," as a subject of research, in order to highlight blind spots in the urban social fabric and social dynamics. Critical urban scholars point to the variability of space and place that cannot be grasped by a binary categorization of "urban" and "rural" or a scalar categorization of cities, which is often performed in empirical research as a consequence of empirical pragmatism or funding logics: "Rather than fully subsuming the rural into the urban, we must strive for deeper understanding of the urban's many constitutive outsides, including the rural, because they are what allow for radical alterity and radical undecidability" (Reddy, 2018, p. 532; see also Glick Schiller & Çağlar, 2011). Thus, recent urban theory suggests a relational and procedural perspective on the urban and its social order, integrating time as relevant research category. In the context of research on forced migrants, one relevant topic for future research could be the comparative analysis of neighborhood relations and the question how they are constituted in time and space, considering variances with regard to ethnicity, race, gender, age, and family or social status, but also with regard to the constituency of space and place. In this context, we must reconsider what scale is meant by the "urban": at what size does this very quality of innovation, transition, and freedom that is commonly devoted to the "urban" begin – and where does it end? Should we differentiate between metropolises, large cities, and small towns when we talk about urban asylum? How can we arrive at a realistic picture of arrival of forced migrants and their contribution to the transformation of urban space? With regard to the numerous case studies and comparative research projects on forced migrant arrival, a meta study and further comparative urban research could help identify relevant aspects, could create a scalar typology of urbanity, and could frame this in relation to "urbanism." This would support connecting global processes of forced migration and urbanization, and relating them to the very meaning of space and place – not just as a spatial formation or structure within the boundaries of what is seen as "the city," but as a quality of social relations (see Darling, 2020, this issue).

This brings us to our last point, which derives from our own position as migration and urban scholars located in the Global North: While we are aware of the great variety of forced migration dynamics, internal and protracted displacement, and refugee reception processes worldwide on the one hand, and on the other, the diversity of urban structures, urbanization processes and migrant urbanism, we also acknowledge the potential limitations of the perspectives offered in this introduction. However, we hope that these perspectives may nonetheless contribute to ongoing debates about the urban as a place of asylum, and to inspire forced migration and urban studies scholars to engage in global comparative, mixed-method and interdisciplinary research and thus reveal global differentiations, but also commonalities and interrelations – in and between both the Global North and the Global South – in order to make progress and rethink some gridlocked discussions in this important research field.

## Note

1. Since the mid-1990s, there has been a highly contested scholarly shift away from "refugee studies" in favor of "forced migration studies," with the attempt to anticipate "knowledge about *all forms* of displacement and its subjects within a transformed international system" (Chimni, 2009, p. 17f.). For detailed analyses and debates about this turn, and the specific epistemologies of both fields, see also Turton (2003).

## Disclosure statement

No potential conflict of interest was reported by the author(s).

## ORCID

René Kreichauf http://orcid.org/0000-0003-0004-6749
Birgit Glorius http://orcid.org/0000-0001-5565-0105

## References

Agier, M., Nice, R., & Wacquant, L. (2002). Between war and city: Towards an urban anthropology of refugee camps. *Ethnography*, 3(3), 317–341. https://doi.org/10.1177/146613802401092779

Agustin, Ó.G., & Jørgensen, M. B. (2019). *Solidarity and the 'Refugee crisis' in Europe*. Palgrave McMillan.

Ambrosini, M. (2021). The battleground of asylum and immigration policies: A conceptual inquiry. *Ethnic and Racial Studies*, 44(3), 374–395. https://doi.org/10.1080/01419870.2020.1836380

Bastide, L. (2014). Globalizing Kuala Lumpur: Indonesian migrant workers, urban borderscapes and the production of metropolitan spaces N., Aveline-Dubach, S.C., Jou, and M., Hsin-Huang. In *Globalization and New Intra-Urban dynamics in Asian cities* (pp. 377–405). National Taiwan University Press.

Baubock, R. (2003). Reinventing urban citizenship. *Citizenship Studies*, 7(2), 139–160. https://doi.org/10.1080/1362102032000065946

Bauder, H. (2017). Sanctuary cities: Policies and practices in international perspective. *International Migration*, 55(2), 174–187. https://doi.org/10.1111/imig.12308

Bazurli, R. (2019). Local governments and social movements in the 'Refugee crisis': Milan and Barcelona as 'Cities of Welcome'. *South European Society and Politics*, 24(3), 343–370. https://doi.org/10.1080/13608746.2019.1637598

Belloni, M. (2016). Learning how to squat: Cooperation and conflict between refugees and natives in Rome. *Journal of Refugee Studies*, 29(4), 506–527. https://doi.org/10.1093/jrs/few033

Bhagat, A. (2020). Governing refugees in raced markets: Displacement and disposability from Europe's frontier to the streets of Paris. *Review of International Political Economy*, 1–24. https://doi.org/10.1080/09692290.2020.1844781

Bose, P.S. (2020). Refugees and the transforming landscapes of small cities in the US. *Urban Geography*, 1–21. https://doi.org/10.1080/02723638.2020.1785202

Brenner, N. (2018). Debating planetary urbanization: For an engaged pluralism. *Environment and Planning D: Society and Space*, 36(3), 570–590. https://doi.org/10.1177/0263775818757510

Brettell, C. B. (2003). Bringing the city back in: Cities as contexts for immigrant incorporation. In N. Foner (Ed.), *American arrivals: Anthropology engages the new immigration* (pp. 163–196). School of American Research Press.

Brickell, K., & Datta, A. (2011). *Translocal Geographies. Spaces, Places, Connections* Routledge.

Campesi, G. (2015). Hindering the deportation machine: An ethnography of power and resistance in immigration detention. *Punishment & Society*, 17(4), 427–453. https://doi.org/10.1177/1462474515603804

Casati, N. (2018). How cities shape refugee centres: 'deservingness' and 'good aid' in a Sicilian town. *Journal of Ethnic and Migration Studies*, 44(5), 792–808. https://doi.org/10.1080/1369183X.2017.1354689

Castles, S., & Miller, M.J. (2014). *The age of migration. International population movements in the modern world.* Macmillan Press.

Chimni, B.S. (2009). The birth of a 'Discipline': From refugee to forced migration studies. *Journal of Refugee Studies*, 22(1), 11–29. https://doi.org/10.1093/jrs/fen051

Crawley, H., & Skleparis, D. (2018). Refugees, migrants, neither, both: Categorical fetishism and the politics of bounding in Europe's 'migration crisis'. *Journal of Ethnic and Migration Studies*, 44(1), 48–64. https://doi.org/10.1080/1369183X.2017.1348224

Danewid, I. (2020). The fire this time: Grenfell, racial capitalism and the urbanisation of empire. *European Journal of International Relations*, 26(1), 289–313. https://doi.org/10.1177/1354066119858388

Darling, J. (2017). Forced migration and the city. Irregularity, informality, and the politics of presence. *Progress in Human Geography*, 41(2), 178–198. https://doi.org/10.1177/0309132516629004

Darling, J. (2020). Refugee urbanism: Seeing asylum "like a city". *Urban Geography*, 1–21. https://doi.org/10.1080/02723638.2020.1763611

De Genova, N. (2013). Spectacles of migrant 'illegality': The scene of exclusion, the obscene of inclusion. *Ethnic and Racial Studies*, 36(7), 1180–1198. https://doi.org/10.1080/01419870.2013.783710

De Hoon, M, Vink, M, & Schmeets, H. (2021). On the move again? Residential trajectories of refugees after obtaining asylum in the Netherlands. *Population Space Place*, 27(2), e2386. https://doi.org/10.1002/psp.2386

Dempsey, K. E. (2021). Migrant agency and counter-hegemonic efforts among asylum seekers in the Netherlands in response to geopolitical control and exclusion. *Geopolitics*, 1–22. https://doi.org/10.1080/14650045.2021.1912019

Du Bois, W. E. B. (1899). The Philadelphia Negro: *A Social Study*. University of Pennsylvania Press.

El Moussawi, H., & Schuermans, N. (2021). From asylum to post-arrival geographies: Syrian and Iraqi refugees in Belgium. *Tijdschrift Voor Economische En Sociale Geografie*, 112(2), 164–178. https://doi.org/10.1111/tesg.12469

Elliott-Cooper, A., Hubbard, P., & Lees, L. (2020). Moving beyond Marcuse: Gentrification, displacement and the violence of un-homing. *Progress in Human Geography*, 44(3), 492–509. https://doi.org/10.1177/0309132519830511

Fawaz, M. (2016a). *Unsettling the "refugee crisis": Notes from Beirut*. International Journal of Urban and Regional Research [WWW essay]. http://www.ijurr.org/spotlight-on-overview/spotlight-urban-refugee-crisis/unsettling-refugee-crisis-notes-beirut/

Fawaz, M. (2016b). Planning and the refugee crisis: Informality as a framework of analysis and reflection. *Planning Theory*, 16(1), 99–115. https://doi.org/10.1177/1473095216647722

Fernandez, D. (2007). *Targeted. National security and the business of immigration*. Seven Stories Press.

Fiddian-Qasmiyeh, E., Loescher, G., Long, K., & Sigona, N. (eds). (2014). *The Oxford Handbook of refugee and forced migration studies*. Oxford University Press.

Fontanari, E. (2015). Confined to the threshold. *City*, 19(5), 714–726. https://doi.org/10.1080/13604813.2015.1071112
Georgi, F. (2019). Toward fortress capitalism: The restrictive transformation of migration and border regimes as a reaction to the capitalist multicrisis. *Canadian Review of Sociology*, 56(4), 556–579. https://doi.org/10.1111/cars.12264
Ghorashi, H., De Boer, M., & Ten Holder, F. (2018). Unexpected agency on the threshold: Asylum seekers narrating from an asylum seeker centre. *Current Sociology*, 66(3), 373–391. https://doi.org/10.1177/0011392117703766
Giddens, A. (1995). *Nation-state and violence*. University of California Press.
Glick Schiller, N., & Çağlar, A. (2011). Downscaled cities and migrant pathways. Locality and agency without an ethnic lens. In N. Glick Schiller & A. Çağlar (Eds.), *Locating migration. Rescaling cities and migrants* (pp. 190–212). Cornell University Press.
Glorius, B., Bürer, M., & Schneider, H. (2021). Integration of refugees in rural areas and the role of the receiving society: Conceptual review and analytical framework. *Erdkunde*, 75(1), 51–60. https://doi.org/10.3112/erdkunde.2021.01.04
Glorius, B., & Doomernik, J. (2020). Introduction. In B. Glorius & J. Doomernik (Eds.), *Geographies of asylum in Europe and the role of European localities* (pp. 1–11). IMISCOE Research Series. Springer Nature.
Gottdiener, M., & Budd, L. (2005). *Key concepts in urban studies*. Key Concepts Series. Sage Publications Ltd.
Grabska, K. (2006). Marginalization in urban spaces of the Global South: Urban refugees in Cairo. *Journal of Refugee Studies*, 19(3), 287–307. https://doi.org/10.1093/jrs/fel014
Grbac, P. (2013). *Civitas, polis, and urbs – Reimagining the refugee camp as the city* (Refugee Studies Centre, Working Paper Series 96, University of Oxford).
Greenwood, M. J., & Hunt, G. L. (2003). The early history of migration research. *International Regional Science Review*, 26(1), 3–37. https://doi.org/10.1177/0160017602238983
Guiraudon, V., & Lahav, G. (2000). A reappraisal of the state sovereignty debate: The case of migration control. *Comparative Political Studies*, 33(2), 163–195. https://doi.org/10.1177/0010414000033002001
Haddad, E. (2002). The refugee: Forging national identities. *Studies in Ethnicity and Nationalism*, 2 (2), 23–38. https://doi.org/10.1111/j.1754-9469.2002.tb00025.x
Hall, S. M. (2015). Migrant urbanisms: Ordinary cities and everyday resistance. *Sociology*, 49(5), 853–869. https://doi.org/10.1177/0038038515586680
Hamann, U., & El-Kayed, N. (2018). Refugees' access to housing and residency in German cities: Internal border regimes and their local variations. *Social Inclusion*, 6(1), 135–146. https://doi.org/10.17645/si.v6i1.1334
Hamlin, R. (2014). *Let me be a refugee: Administrative justice and the politics of asylum in the United States, Canada, and Australia*. Oxford University Press.
Hartmann, M. (2017). Contested boundaries: Refugee centers as spaces of the political. *Z'Flucht. Zeitschrift für Flucht-und Flüchtlingsforschung*, 1(2), 218–243. https://doi.org/10.5771/2509-9485-2017-2-218
Harvey, D. (1996). *Justice, nature, and the geography of difference*. Blackwell Publishers.
Harvey, D. (2000). *Possible urban worlds, megacities lecture (4)*. Twynstra Gudde Management Consultants.
Hess, S. (2015). Politiken der (Un-)Sichtbarmachung. Eine Kritik der Wissens- und Bilderproduktionen zu Migration. In E. Yildiz & M. Hill (Eds.), *Nach der Migration. Postmigrantische Perspektiven jenseits der Parallelgesellschaft* (pp. 49–64). transcript.
Hill, E., Meer, N., & Peace, T. (2021). The role of asylum in processes of urban gentrification. *The Sociological Review*, 69(2), 259–276. https://doi.org/10.1177/0038026120970359
Hooghe, L., & Marks, G. (2001). *Multi-level governance and European integration*. Rowman & Littlefield.
Horvath, K., Amelina, A., & Peters, K. (2017). Re-thinking the politics of migration. On the uses and challenges of regime perspectives for migration research. *Migration Studies*, 5(3), 301–314. https://doi.org/10.1093/migration/mnx055

Huq, E., & Miraftab, F. (2020). "We are all refugees": Camps and informal settlements as converging spaces of global displacements. *Planning Theory & Practice*, 21(3), 351–370. https://doi.org/10.1080/14649357.2020.1776376

Hyndman, J. (2000). *Managing displacement: Refugees and the politics of humanitarianism*. University of Minnesota Press.

Ikizoglu Erensu, A., & Kaşli, Z. (2016). A tale of two cities: Multiple practices of bordering and degrees of 'Transit' in and through Turkey. *Journal of Refugee Studies*, 29(4), 528–548. https://doi.org/10.1093/jrs/few037

Jansen, B. J. (2016 The protracted refugee camp and the consolidation of a 'humanitarian urbanism'). International Journal of Urban and Regional Research [WWW essay]. https://www.ijurr.org/spotlight-on/the-urban-refugee-crisis-reflections-on-cities-citizenship-and-the-displaced/the-protracted-refugee-camp-and-the-consolidation-of-a-humanitarian-urbanism/]

Jewkes, Y. (2013). On carceral space and agency. In D Moran, N Gill, & D Conlon (Eds.), *Carceral spaces: Mobility and agency in imprisonment and migrant detention* (pp. 127–132). Ashgate Publishers.

Kandylis, G. (2019). Accommodation as displacement: Notes from refugee camps in Greece in 2016. *Journal of Refugee Studies*, 32(1), i12–i21. https://doi.org/10.1093/jrs/fey065

Katz, I., Parsloe, T., Poll, Z., & Scafe-Smith, A. (2018). The bubble, the airport, and the jungle: Europe's urban migrant camps. In I. Katz, D. Martin, & C. Minca (Eds.), *Camps revisited: Multifaceted spatialities of a modern political technology* (pp. 61–82). Rowman & Littlefield.

Koçak, M. (2020). Who is "Queerer" and deserves resettlement?: Queer asylum seekers and their deservingness of refugee status in Turkey. *Middle East Critique*, 29(1), 29–46. https://doi.org/10.1080/19436149.2020.1704506

Kos, S., Maussen, M., & Doomernik, J. (2016). Policies of exclusion and practices of inclusion: How municipal governments negotiate asylum policies in the Netherlands. *Territory, Politics, Governance*, 4(3), 354–374. https://doi.org/10.1080/21622671.2015.1024719

Koser, K., & Martin, S. (Eds.). (2011). *The migration-displacement nexus: Patterns, processes, and policies*. Berghahn Books.

Kreichauf, R., & Mayer, M. (2021). Negotiating urban solidarities: Multiple agencies and contested meanings in the making of solidarity cities. *Urban Geography*, 1–24. https://doi.org/10.1080/02723638.2021.1890953

Kreichauf, R. (2021a). Detention as social space: Waiting, social relations, and mundane resistance of asylum seekers in detention. *Critical Sociology*, 47(4–5), 745–762. https://doi.org/10.1177/0896920520977644

Kreichauf, R. (2021b). Forced migration and legal-spatial violence in the United States and Germany. *Political Geography*, 87(2021), 102371. https://doi.org/10.1016/j.polgeo.2021.102371

Kron, Stefanie, & Lebuhn, Henrik. (2020). Building solidarity cities. From protest to policy. In Feyzi Baban & Kim Rygiel (Eds.), *Living with others. Fostering cultural pluralism through citizenship politics* (pp. 81–105). Routledge.

Kühn, M. (2018). Immigration strategies of cities: Local growth policies and urban planning in Germany. *European Planning Studies*, 26(9), 1747–1762. https://doi.org/10.1080/09654313.2018.1484428

Landry, O. (2015). "Wir sind alle oranienplatz"! Space for refugees and social justice in Berlin. *Eminar: A Journal of Germanic Studies*, 51(4), 398–413. https://doi.org/10.3138/seminar.2015.51.4.398

Lebuhn, H. (2013). Local border practices and urban citizenship in Europe: Exploring urban borderlands. *CITY*, 17(1), 37–51. https://doi.org/10.1080/13604813.2012.734072

Levitt, P., DeWind, J., & Vertovec, S. (2003). International perspectives on transnational migration: An introduction. *International Migration Review*, 37(3), 565–575. https://doi.org/10.1111/j.1747-7379.2003.tb00150.x

Lidén, G., & Nyhlén, J. (2015). Reception of refugees in Swedish municipalities: Evidences from comparative case studies. *Migration and Development*, 4(1), 55–71. https://doi.org/10.1080/21632324.2014.894178

Lyytinen, E. (2015). Congolese refugees' 'right to the city' and urban (in)security in Kampala, Uganda. *Journal of Eastern African Studies*, 9(4), 593–611. https://doi.org/10.1080/17531055.2015.1116142

Malkki, L. H. (1995). Refugees and exile: From 'Refugee studies' to the national order of things. *Annual Review of Anthropology*, 24(1), 495–523. https://doi.org/10.1146/annurev.an.24.100195.002431

Malkki, L. H. (2002). News from nowhere. *Ethnography*, 3(3), 351–360. https://doi.org/10.1177/146613802401092797

Manfredi-Sánchez, J. L. (2020). Sanctuary cities: What global migration means for local governments. *Social Sciences*, 9(8), 146, 1–16. https://doi.org/10.3390/socsci9080146

Marchetti, C. (2020). Cities of exclusion: Are local authorities refusing asylum seekers? In M. Ambrosini, M. Cinalli, & D. Jacobson (Eds.), *Migration, borders and citizenship. Migration, diasporas and citizenship* (pp. 237–263). Palgrave Macmillan.

Marcuse, P. (2013). *Abandonment, gentrification, and displacement: The linkages in New York City (169-193)*. Routledge.

Martin, D., Minca, C., & Katz, I. (2020). Rethinking the camp: On spatial technologies of power and resistance. *Progress in Human Geography*, 44(4), 743–768. https://doi.org/10.1177/0309132519856702

Martin, D. (2015). From spaces of exception to 'campscapes': Palestinian refugee camps and informal settlements in Beirut. *Political Geography*, 44(January2015), 9–18. https://doi.org/10.1016/j.polgeo.2014.08.001

Massey, D. S. (1999). International migration at the Dawn of the twenty-first century: The role of the state. *Population and Development Review*, 25(2), 303–322. https://doi.org/10.1111/j.1728-4457.1999.00303.x

Mayer, M. (2018). Cities as sites of refuge and resistance. *European Urban and Regional Studies*, 25 (3), 232–249. https://doi.org/10.1177/0969776417729963

Meeus, B., Van Heur, B., & Arnaut, K. (2019). *Arrival infrastructures. Migration and urban social mobility*. Palgrave Macmillan.

Meret, S., & Diener, W. (2019). We are still here and staying! Refugee-led mobilizations and their struggles for rights in Germany. In B. Siim, A. Krasteva, & A. Saarinen (Eds.), *Citizens' activism and solidarity movements* (pp. 137–166). Palgrave Macmillan.

Moulier Boutang, Y. (2007). Europa, Autonomie der Migration, Biopolitik. In M. Pieper, K. Serhat Thomas, & T. Vassilis (Eds.), *Empire und die biopolitische Wende. Die Internationale Diskussion im Anschluss an Negri und Hardt* (pp. 169–180). Campus.

Muggah, R., & Abdenur, A.E. (2018). *Refugees* and the City. The Twenty-f*irst-century front line* (World Refugee Council Research Paper No. 2). Centre for International Governance Innovation.

Müller-Funk, L. (2020). Research with refugees in fragile political contexts: How ethical reflections impact methodological choices. *Journal of Refugee Studies*, 34(2), 2308–2332. https://doi.org/10.1093/jrs/feaa013

Myrberg, G. (2017). Local challenges and national concerns: Municipal level responses to national refugee settlement policies in Denmark and Sweden. *International Review of Administrative Sciences*, 83(2), 322–339. https://doi.org/10.1177/0020852315586309

Neis, H. J., Meier, B., & Furukawazono, T. (2018). Welcome city: Refugees in three German cities. *Urban Planning*, 3(4), 101–115. https://doi.org/10.17645/up.v3i4.1668

Nettelbladt, G., & Boano, C. (2019). Infrastructures of reception: The spatial politics of refuge in Mannheim, Germany. *Political Geography*, 71(May2019), 78–90. https://doi.org/10.1016/j.polgeo.2019.02.007

Oomen, B. (2020). Decoupling and teaming up: The rise and proliferation of transnational municipal networks in the field of migration. *International Migration Review*, 54(3), 913–939. https://doi.org/10.1177/0197918319881118

Papadopoulos, S., & Tsianos, V.S. (2013). After citizenship: Autonomy of migration, organisational ontology and mobile commons. *Citizenship Studies*, 17(2), 178–196. https://doi.org/10.1080/13621025.2013.780736

Phillimore, J., & Goodson, L. (2006). Problem or opportunity? Asylum seekers, refugees, employment and social exclusion in deprived urban areas. *Urban Studies*, 43(10), 1715–1736. https://doi.org/10.1080/00420980600838606

Picker, G., & Pasquetti, S. (2015). Durable camps: The state, the urban, the everyday. *City*, 19(5), 681–688. https://doi.org/10.1080/13604813.2015.1071122

Porter, L., Sanyal, R., Bergby, S., Yotebieng, K., Lebuhn, H., Ramírez, M M., Neto, P F., & Tulumello, S. (2019). Introduction: Urban planning and the global movement of people. *Planning Theory & Practice*, 20(1), 99–128. https://doi.org/10.1080/14649357.2018.1558566

Pottie-Sherman, Y. (2018). Austerity urbanism and the promise of immigrant- and refugee-centered urban revitalization in the US Rust Belt. *Urban Geography*, 39(3), 438–457. https://doi.org/10.1080/02723638.2017.1342398

Rajaram, P.K. (2018). Refugees as surplus population: Race, migration and capitalist value regimes. *New Political Economy*, 23(5), 627–639. https://doi.org/10.1080/13563467.2017.1417372

Ramadan, A. (2013). Spatialising the refugee camp. *Transactions of the Institute of British Geographers*, 38(1), 65–77. https://doi.org/10.1111/j.1475-5661.2012.00509.x

Ravenstein, E.G. (1885). The laws of migration – I. *Journal of the Statistical Society*, 48(2), 167–227. https://doi.org/10.2307/2979181

Ravenstein, E.G. (1889). The laws of migration – II. *Journal of the Statistical Society*, 52(2), 214–301. https://doi.org/10.2307/2979333

Reddy, R.N. (2018). The urban under erasure: Towards a postcolonial critique of planetary urbanization. *Environment and Planning D: Society and Space*, 36(3), 529–539. https://doi.org/10.1177/0263775817744220

Rickards, L., Gleeson, B., Boyle, M., & O'Callaghan, C. (2016). Urban studies after the age of the city. *Urban Studies*, 53(8), 1523–1541. https://doi.org/10.1177/0042098016640640

Robertson, S. (2014). Migrants as scale makers: Untangling the intersections of urban theory and migration research. *City*, 18(2), 230–233. https://doi.org/10.1080/13604813.2014.896656

Römhild, R. (2015). Jenseits ethnischer Grenzen. Für eine postmigrantische Kultur- und Gesellschaftsforschung. In E. Yildiz & M. Hill (Eds.), *Nach der Migration. Postmigrantische Perspektiven jenseits der Parallelgesellschaft* (pp. 37–48). transcript.

Sanyal, R. (2016). *From camps to urban refugees: Reflections on research agendas.* International Journal of Urban and Regional Research [WWW essay]. http://www.ijurr.org/spotlight-on-overview/spotlight-urban-refugee-crisis/camps-urban-refugees-reflections-research-agendas/

Sanyal, R. (2012). Refugees and the city: An urban discussion. *Geography Compass*, 6(11), 633–644. https://doi.org/10.1111/gec3.12010

Sanyal, R. (2014). Urbanizing refuge: Interrogating spaces of displacement. *International Journal of Urban and Regional Research*, 38(2), 558–572. https://doi.org/10.1111/1468-2427.12020

Sanyal, R. (2017). A no-camp policy: Interrogating informal settlements in Lebanon. *Geoforum*, 84, 117–125. https://doi.org/10.1016/j.geoforum.2017.06.011

Sanyal, R. (2021). Making urban humanitarian policy: The "neighbourhood approach" in Lebanon. *Urban Geography*, 1–21. https://doi.org/10.1080/02723638.2020.1849983

Sassen, S. (2013). When the center no longer holds: Cities as frontier zones. *Cities*, 34 (October2013), 67–70. https://doi.org/10.1016/j.cities.2012.05.007

Sassen, S. (2014). *Expulsions. Brutality and complexity in the global economy.* Harvard University Press.

Schmidtke, O. (2014). Beyond national models? Governing migration and integration at the regional and local levels in Canada and Germany. *Comparative Migration Studies*, 1(1), 77–99. https://doi.org/10.5117/CMS2014.1.SCHM

Scholten, P., & Penninx, R. (2016). The multilevel governance of migration and integration. In Garcés-Mascareñas, R. Penninx (Ed.), *Integration processes and policies in Europe contexts, levels and actors* (pp. 91–108). Springer.

Schuster, L. (2005). A Sledgehammer to crack a nut: Deportation, detention and dispersal in Europe. *Social Policy and Administration*, 39(6), 606–621. https://doi.org/10.1111/j.1467-9515.2005.00459.x

Seethaler-Wari, S. (2018). Urban planning for the integration of refugees: The importance of local factors. *Urban Planning*, 3(4), 141–155. https://doi.org/10.17645/up.v3i4.1696
Semprebon, M. (2021). Towards a parallel exceptional welfare system: The scaling down and out of forced migrants' reception in Italy. *Urban Geography*, 1–22. https://doi.org/10.1080/02723638.2021.1890424
Sigona, N. (2015). Campzenship: Reimagining the camp as a social and political space. *Citizenship Studies*, 19(1), 1–15. https://doi.org/10.1080/13621025.2014.937643
Simmel, G. (1903/2012). *The metropolis and mental life*. Routledge.
Simmel, G. (1908/2008). *The stranger*. Routledge.
Simone, A.M.M. (2004). People as infrastructure: Intersecting fragments in Johannesburg. *Public Culture*, 16(3), 407–429. https://doi.org/10.1215/08992363-16-3-407
Smith, M. P. (2001). *Transnational urbanism: Locating globalization*. Blackwell.
Soederberg, S. (2018). Governing global displacement in austerity urbanism: The case of Berlin's refugee housing crisis. *Development and Change*, 50(4), 923–947. https://doi.org/10.1111/dech.12455
Steiner, H., & Reinhard, A. (2018). A culture of welcome: Challenges and experiences in the city of Freiburg/Germany. In H. Kury & S. Redo (Eds.), *Refugees and migrants in law and policy* (pp. 379–391). Springer.
Suerbaum, M. (2020). *Masculinities and displacement in the Middle East: Syrian refugees in Egypt*. I.B. Tauris.
Szczepanikova, A. (2013). Between control and assistance: The problem of European accommodation centres for asylum seekers. *International Migration*, 51(4), 130–131. https://doi.org/10.1111/imig.12031
Tazzioli, M., Garelli, G., De Genova, N., De Genova, N., Garelli, G., & Tazzioli, M. (2018). Autonomy of asylum?: The autonomy of migration undoing the refugee crisis script. *South Atlantic Quarterly*, 117(2), 239–265. https://doi.org/10.1215/00382876-4374823
Tramonte, L. (2011). *Debunking the Myth of sanctuary cities. Community policing policies protect Americans* (Special Report). Immigration Policy Center.
Turner, S. (2016). What is a refugee camp? Explorations of the limits and effects of the camp. *Journal of Refugee Studies*, 29(2), 139–148. https://doi.org/10.1093/jrs/fev024
Turton, D. (2003). *Conceptualising forced migration* (RSC Working Paper, 12). University of Oxford.
Turton, D. (2005). The meaning of place in a world of movement: Lessons from long-term field research in Southern Ethiopia. *Journal of Refugee Studies*, 18(3), 258–280. https://doi.org/10.1093/refuge/fei031
Urry, J. (2007). *Mobilities*. Polity.
Van Liempt, I., & Miellet, S. (2021). Being far away from what you need: The impact of dispersal on resettled refugees' homemaking and place attachment in small to medium-sized towns in the Netherlands. *Journal of Ethnic and Migration Studies*, 47(11), 2377–2395. https://doi.org/10.1080/1369183X.2020.1845130
Van Vliet, J., Eitelberg, D. A., & Verburg, P. H. (2017). A global analysis of land take in cropland areas and production displacement from urbanization. *Global Environmental Change*, 43 (March 2017) , 107–115. https://doi.org/10.1016/j.gloenvcha.2017.02.001
Varsanyi, M.W. (2006). Interrogating "Urban citizenship" vis-à-vis undocumented migration. *Citizenship Studies*, 10(2), 229–249. https://doi.org/10.1080/13621020600633168
Watson, J. (2019). Welcoming refugees and the cultural wealth of cities: Intersections of urban development and refugee humanitarianism. *International Journal of Urban and Regional Research*, 43(5), 983–999. https://doi.org/10.1111/1468-2427.12736
Weidinger, T., & Kordel, S. (2020). Access to and exclusion from housing over time: Refugees' experiences in rural areas. *International Migration, Epub Ahead of Print*. https://doi.org/10.1111/imig.12807
Werner, F., Haase, A., Renner, N., Rink, D., Rottwinkel, M., & Schmidt, A. (2018). The local governance of arrival in Leipzig: Housing of asylum-seeking persons as a contested field. *Urban Planning*, 3(4), 116–127. https://doi.org/10.17645/up.v3i4.1708

Wimmer, A., & Glick Schiller, N. (2002). Methodological nationalism and beyond: Nation–state building, migration and the social sciences. *Global Networks*, 2(4), 301–334. https://doi.org/10.1111/1471-0374.00043

Wren, K. (2003). Refugee dispersal in Denmark: From macro- to micro-scale analysis. *International Journal of Population Geography*, 9(1), 57–75. https://doi.org/10.1002/ijpg.273

Yiftachel, O. (2020). From displacement to displaceability. *City*, 24(1–2), 151–165. https://doi.org/10.1080/13604813.2020.1739933

Yildiz, E. (2014). *Die Weltoffene Stadt*. transcript.

Yurdakul, G., & Korteweg, A. (2021). Boundary Regimes and the Gendered Racialized Production of Muslim Masculinities: Cases from Canada and Germany. *Journal of Immigrant and Refugee Studies*, 19(1), 39–54. https://doi.org/10.1080/15562948.2020.1833271

Zapata-Barrero, R., Caponio, T., & Scholten, P. (2017). Theorizing the 'local turn' in a multilevel governance framework of analysis: A case study in immigrant policy. *International Review of Administrative Sciences*, 83(2), 241–246. https://doi.org/10.1177/0020852316688426

Zetter, R. (2015, September 10). *Protection in crisis: Forced migration and protection in a global era*. Migration Policy Institute. www.migrationpolicy.org/sites/default/files/publications/TCM-Protection-Zetter.pdf

Žižek, S. (2016). *Against the double blackmail: Refugees, terror and other troubles with the neighbours*. Penguin Books.

OPEN ACCESS

# Refugee urbanism: seeing asylum "like a city"

Jonathan Darling

**ABSTRACT**
This paper explores how asylum might be understood from an urban perspective. The paper focuses on a range of conceptual interventions mobilized around the notion of 'seeing like a city', which foreground the pragmatic and compositional nature of urban politics. These debates are placed in conversation with discussions over the changing nature of relations of power, to examine the dispersal of asylum seekers in the UK, arguing that current research on dispersal has tended to focus too exclusively on the regulatory and sovereign aspects of this policy. In doing so, such research has overlooked the urban manifestation of dispersal as a process that has created new knowledges and forms of expertise, whilst being sustained through spatially complex relations of present authority and distant accountability. In exploring these relations, the paper argues that a focus on 'refugee urbanism' may open productive new avenues for critically exploring urban asylum.

It's 2014 and I'm sitting in a café talking with Anna, an asylum seeker originally from Eastern Europe, who has lived in Glasgow for two years. She is accommodated as part of the UK's policy of dispersing asylum seekers across the country. We're discussing the housing conditions she and her two children have faced. She tells me; "I have changed house three times, because I always had problems. Sometimes with the boiler, very bad conditions, but you can't call Scottish Gas, you need to call Orchard and Shipman. This Orchard and Shipman are doing nothing … then I had problems with the roof, because water started to leak from the roof and I started to call them, call them, and no one come and so I moved my children to another room, because I was worried. And one day it collapsed, the roof collapsed, it was impossible to stay there and they told me that urgently we need to move. You can't choose where you are to move, and they moved me at the other end of Glasgow. We suffered at this place … we knew no one and had no idea where to go. I was calling, have you repaired, have you repaired? No, no, no, no … after five months of me calling, of Refugee Council calling, Home Office, Serco, Orchard and Shipman, they repaired and I was very happy. They told me leave your stuff because you will be returned. But five months I need some stuff. So I took a taxi … I open the door and what I see, someone else lives there and I lost my stuff".

Anna's words reflect many of the disciplinary attributes often associated with dispersal: the enforced mobility to Glasgow; the disruption to social networks of a move across the city without any choice; the frustrations of poor housing conditions; and a lack of response to complaints. In this sense, the experiences of Anna and her family form part of a repeated pattern of ill-treatment of those seeking asylum in the UK (see Home Affairs Committee, 2018; National Audit Office, 2014; Scottish Refugee Council, 2014). Read in this context, Anna's account is one among many. This account emerged during research that explored the relationship between asylum seekers and cities in the UK, centered around the changing nature of dispersal as a policy to accommodate asylum seekers (Darling, 2016a). Anna's account clearly speaks to a story of asylum seekers accommodated as a means of regulation and control, and subject to inadequate and at times unsanitary conditions. In this reading, asylum seekers are managed, monitored, and controlled according to conditions defined by the sovereign authority of the Home Office, an authority actualized in and through cities like Glasgow.

Whilst Anna's account reflects the disciplinary role of dispersal, this is only one reading, for her account also presents a more complicated engagement with the city. Beyond the sovereign imposition of dispersal, we might consider the multiple authorities at play in mediating accommodation, in lodging and responding to complaints, and in advocating for Anna's rights. We might also attend to the socio-material infrastructures that frame Anna's experiences of security in the city, and the materials that compose and deconstruct her flat – from the inadequate heating system, to the bulging ceiling as water accumulates. Anna's experience goes beyond an image of dispersal as a regulatory device and of asylum as a matter of sovereign authority alone. Rather, it reflects asylum when read through the complexity of how urban worlds are composed, contested, and conceptualized. In this paper, I explore how we might examine these urban connections as a means to disentangle the relations between "the claims of sovereign power" and "the *techniques* and *projects* that seek to actualize those claims" (Dunn & Cons, 2014, p. 105).

Recent years have seen a growth of work examining the relationship between refugees and cities in the global North. This trend might be argued to reflect two connected factors. First, the so-called "refugee crisis" in Europe has led to growing numbers of asylum seekers and refugees in cities across the global North, prompting a range of critical accounts of urban accommodation, regulation, and rights claims (Depraetere & Oosterlynck, 2017; Koptyaeva, 2017; Kreichauf, 2018; Nordling et al., 2017). Second, there has been an emergent interest in whether theories of urban diversity reflect the specificities of how asylum seekers and refugees interact with cities (Darling, 2017). This demands a move beyond generalized accounts of urban migration patterns, to situate the specificities of refugee subjectivity in relation to a range of urban issues, including housing informality (Sanyal, 2014), racialized inequality (Lancione, 2016; Picker, 2017), contestations over urban resources (Maestri, 2017), and the rise of urban sanctuary movements (Bagelman, 2016; Bauder, 2017; Darling and Bauder 2019).

It is my contention that despite growing attention, these discussions continue to lack conceptual frameworks within which to explore urban refuge. This is because they often begin from a perspective of examining *either* refugee subjectivities and forms of agency, *or* examining urban forms of governance, policy, and claims-making. The intersection of these points of governance, agency, subjectivity and urbanity, is rarely at the forefront of analysis, and it is this configuration that I develop in using the term "refugee urbanism". Refugee urbanism is not intended to imply a generic urban refugee experience, but rather to point to the conjuncture of factors that shape how refugee subjectivities intersect with, are remade by, and themselves serve to constitute and rework, the urban. Refugee urbanism denotes an analytic configuration that goes beyond the study of "urban refugees", as this work often prioritizes one or other of its constituent terms (Darling, 2017). Refugee urbanism asserts that the intersection of such terms matters, both conceptually and politically. In proposing an approach focused on refugee urbanism, this paper offers one provisional framework for studying the political contestations of asylum through the specific politics, conditions, and constraints of the urban. To do so, I focus on a body of work that has mobilized the notion of "seeing like a city" as a means to understand and explore urbanism as a way of life (Amin & Thrift, 2017; Valverde, 2011; Magnusson, 2011). This work challenges assumptions of singular or coherent sovereign authority, and views politics, from the perspective of the city, as entailing not only a multiplicity of authorities vying for influence, but also the necessity of negotiation among and between these varied authorities. Seeing like a city names an approach that accounts for the restlessness of urban change, with orders being perceived as moments of brief stabilization within conditions of flux. Building on this approach, I argue that seeing like a city might be productively placed in conversation with recent work in geography exploring the changing spatiality of power relations (Allen, 2016). I, therefore, argue that advancing an approach to urban asylum orientated around seeing like a city, may better apprehend the complexities of how government is both enacted and contested.

In considering what seeing like a city may offer to an understanding of urban asylum, the paper draws on an empirical project exploring the relationship between asylum seekers, dispersal and urban governance in the UK (Darling, 2016a). Focused on four cities (Birmingham, Cardiff, Glasgow and Sunderland), this project involved 105 interviews with stakeholders in the asylum accommodation and support sector to critically examine the changing nature of dispersal and its effects on refugees and asylum seekers. These interviews included local authority representatives, councillors, service providers, support organizations, third sector groups, refugees and asylum seekers. Empirically, this paper focuses on negotiations around the provision of dispersal accommodation and draws on anonymized interviews from the Glasgow fieldwork, together with media reports collated as part of a local media analysis, tracing narratives of dispersal from 2000 to 2015.

The paper develops as follows. I begin by outlining the main elements of Magnusson's (2011) account of "seeing like a city", alongside discussions of the phrase by Amin and Thrift (2017), and Valverde (2011). In each case, seeing like a city highlights an attachment to the complexity of governing the urban. I then develop this framing further through arguing that to "see like a city" entails an engagement with the changing nature of power relations (Allen, 2016), which considers authority as not simply multiple but as expressed in multiple relations of "reach". With this confluence in mind, I explore the dispersal of asylum seekers across Britain. In reflecting on what seeing like a city may offer to an understanding of urban asylum, the paper concludes by discussing what forms of political potential it brings to light.

## Seeing like a city

In discussing recent debates in urban political theory, Rodgers et al. (2014, p. 4), argue that there is a longstanding tension between two theorizations of urban politics;

> on the one hand, there are concerns for a politics that goes on in cities, which is contained at certain scales, or emerges from specific territories and spatial environments; on the other

hand, there are concerns for a looser politics of cities, which might exceed, extend beyond, filter through or problematize particular scalar configurations.

This relation between a politics *in* the city and a politics *of* the city is one that runs through different mobilizations of "seeing like a city". A common aspect of these accounts is to examine how dimensions of urban life - sociotechnical infrastructures in the case of Amin and Thrift (2017), and land use regulations in the case of Valverde (2011) - express both the everyday politics of cities and a wider political ontology with implications for how political practices, relations, and contestations are understood beyond urban debates. This perspective reaches its fullest extent in Magnusson's discussion of the "politics of urbanism".

The starting point for Magnusson's (2011, p. 50) version of seeing like a city is to view the city as a "political production" that is always in the process of being made. In this sense, Magnusson is the most explicit of the authors working with this term in framing "seeing like a city" as a political ontology first and foremost. Magnusson (2011, p. 2) argues that our political tendency is to "imagine things from the viewpoint of a sovereign government or a sovereign people", and thus to "see like a state" after James C. Scott (1998). This is a tendency that is perhaps most prevalent in work on forced migration, where the framework of the nation-state and its claims to sovereignty, security, and the right to delineate citizenship, has understandable prominence despite critique (Gill, 2010; Mountz, 2010). This is a perceptual orientation that attempts to classify and categorize in an effort to ensure coherence, to maintain a hierarchy of authority that is both top-down and neatly arranged. Of course, as Scott (1998) and others have noted, the realities of state projects are far from such an idealized account, yet underpinning such realities remains a set of perceptual assumptions over where authority resides, how it is expressed, and what it seeks to achieve. In this way, Scott argues that seeing like a state seeks to reduce contingency and render life legible, thereby "impoverish[ing] everyday life" (Taylor, 2013, p. 800). In this context, Magnusson's *Politics of Urbanism* is less a book about urban governance, and more an argument for a political ontology that contests a singular or wholly authoritative orientation toward governing.

Magnusson (2011) thus argues for an account of politics that foregrounds multiple claims to authority. Drawing on Foucault's (2003) discussion of government as the ability to shape the conduct of others, Magnusson argues that;

> urbanism implies proximate diversity, complicated patterns of government and self-government, a multiplicity of authorities in different registers, the infinite deferral of sovereignty, self-organization and an emergent order that, though chaotic, is by no means anarchic. (Magnusson, 2014, p. 11)

This account of "urbanism as a way of life", emerges from a long tradition of urban thought that has examined the characteristics of urban forms of life (Lefebvre, 2003; Wirth, 1938). By placing these discussions in conversation with Foucauldian discussions of governmentality, Magnusson argues that it is in the negotiations of urban life that we might see forms of governmentality and rule most vividly expressed, challenged, and experimented with. In considering the city as a site of governmentality, Magnusson advocates a political ontology that reflects contingent forms of governing;

> To see like a city is to accept a certain disorderliness, unpredictability, and multiplicity as inevitable, and to pose the problem of politics in relation to that complexity, rather than in relation to the simplicity that sovereignty seeks. (Magnusson, 2011, p. 120)

The implication is that whilst multiple political authorities interact, they are never able to fully control or fully predict potential challenges or outcomes. "Seeing like a city" entails recognizing the limits of government created by the complexities of urban life, and in response not seeking a simplification of that complexity by imposing upon it a singular lens of authority or legibility. Magnusson's (2011) account of "seeing like a city" is thus one that eschews any straightforward answers to political questions, be they of rights, authority, or identity. Taking such a politics seriously implies a need to stay with the process of constructing temporary orders and examining their stabilization through diverse forms of power in urban life.

A similar sensibility toward the complexities of government is evident in Valverde's (2011) mobilization of "seeing like a city". In examining land use legislation, she argues that a modernist approach to land use did not wholly replace prior modes of regulating urban space. Rather, she demonstrates how alternative logics continue to surface within urban regulations in different ways. Valverde (2011, p. 280) argues that "governing urban disorder through embodied, experiential, and relational categories is a necessary component of contemporary urban governance", indicating not a transition from seeing like a state to seeing like a city, but rather a creative engagement between these political ontologies. Valverde thus highlights that the notions of seeing like a state or a city do not name unwavering forms of government, they name tendencies toward governing arranged around particular assumptions of authority, knowledge, and hierarchy. These forms of governing are grounded in political ontologies that begin either with the state or with the city as their focus of attention. As such, Valverde (2011, p. 281) presents seeing like a city as centered on "the flexible relation between knowledges" in governing. She develops the term to indicate a "pragmatic approach that uses both old and new gazes, premodern and modern knowledge formats, in a nonzero-sum manner and in unpredictable and shifting combinations". Just as Magnusson (2011) argues that seeing like a city better reflects the realities of urbanism as a way of life, so Valverde (2011, p. 291) argues that seeing like a city better captures the pragmatic compromises associated with urban development. The compromised nature of authority in this mobilization of seeing like a city resonates with critical discussions of the improvised nature of governance in contexts of bordering (Burridge et al., 2017; Dunn & Cons, 2014; Sanyal, 2018). For instance, Dunn (2012) highlights how an "adhocracy" of improvised solutions and guesswork shape humanitarian governance in practice, drawing attention to the limits of authority claims, and presenting a model of sovereignty that is "aleatory" in nature (Dunn & Cons, 2014).

Unlike Magnusson, Valverde (2011) is less concerned with expanding the reach of this urban politics to encompass wider patterns of government. Rather, her lens remains focused on techniques of government *within* cities and what these may tell us about evolving systems of urban regulation. Two points are significant here. First, is the pragmatism that Valverde associates with seeing like a city as a political ontology that requires the incorporation, negotiation, and reworking of different frames of perception and governance, and that often produces imperfect compromises and improvisations.

Second, is the aleatory nature of authority that demands pragmatism in the first place. In this way, Valverde's account highlights that it is in the gap between governmental planning and the actualization of such plans, that the realities of urban politics often intervenes (McFarlane, 2011; Zeiderman et al., 2015). This is valuable in the context of refugee urbanism as it draws attention to the "conjunctural" nature of power and its exercise (Dunn & Cons, 2014, p. 102). Focus is thus drawn to how the conjuncture of multiple claims to, and intensities of, authority, lead to "shifting landscapes of unpredictable power with which both the governed and the governing must contend" (ibid, p. 102). In contending with such instability and adhocracy, efforts to establish order and control present opportunities for varying forms of response, as the "overlaps and dissonances between competing modes of sovereignty or projects of rule create navigable channels" for both the governed and the governing (Dunn & Cons, 2014, p. 100). This interplay between "navigable channels" of responding to authority, and efforts at ordering and controlling urban life, offers a further means to examine the temporary stabilizations that are at the heart of seeing like a city as a political ontology. In this vein, seeing like a city represents an entry point into the pragmatism of navigating complexity without submitting to an image of either totalizing control or complete instability.

It is to these gaps that Amin and Thrift (2017, p. 3) orientate their account of seeing like a city, as a means to apprehend "the combined vitality and political economy of urban sociotechnical systems". In exploring the sociotechnical arrangements that sustain life, Amin and Thrift (2017, p. vii) argue that seeing like a city entails recognizing and responding to "the ontology of spatial 'throwntogetherness'" that Massey (2005) suggests is fundamental to relational urbanism (Massey, 2007). Such "throwntogetherness" names a situation of heterogeneous elements juxtaposed to produce relational connections that bring diversity into proximity. Throwntogetherness demands negotiation on multiple levels, from questions of responsibility for relations of implication (Massey, 2007), to the challenges of encountering difference as a disruptive and potentially transformative part of urban life (Wilson, 2017; Darling & Wilson, 2016). In this relational imaginary, seeing like a city is positioned as a means to apprehend the juxtapositions of order and discipline, discretion and disposition, and the material and the affective, which shape how sociotechnical systems influence urban life, and which forge particular types of infrastructural politics. Thus, just as Magnusson and Valverde mobilize seeing like a city as an account of political practice, so Amin and Thrift (2017, p. 120) argue that urban infrastructures are "political in every way: governed in favour of particular interests, biased in their affordances and allocations, shot through with calculative logics and mechanisms designed to distribute unevenly". As a result, political questions, when seen like a city, are centered on how we deal with this complicated, unpredictable, and often shifting sense of order.

There are three implications of this proposition for exploring urban asylum. First, and perhaps most obviously, it implies that it is not productive to rely upon sovereign claims to authority or to seek in them solutions to political challenges. This is not simply because such claims have been most readily associated with the violence of the state, although this is clearly significant when considering the lives of asylum seekers (Gill, 2010). Rather, it is because relying upon sovereign authority risks dismissing a much wider range of claims to authority and influence, which themselves negotiate, mould, and influence claims to sovereignty. Second, it implies that authority is not a zero-sum game, but rather that authority is produced in different collaborations and contexts, and may intensify and diminish at different points. Finally, seeing like a city demands an appreciation of uncertain and unexpected relations of government and authority, of how lives may be governed in ways unanticipated and perhaps unintended by those assumed to be "in control". Not only are urban authorities "constantly responding to problems that they cannot quite control", but in doing so they "interact in ways that lead to unpredictable results" (Magnusson, 2015, pp. 25–26). Read as such, seeing like a city focuses attention on the variety of actors who may be involved in sustaining, and contesting, an imaginary of sovereign authority.

## Powers of reach

Before moving to consider the application of this approach, there is a need to consider the complexities of how authority is spatially produced. This entails a concern with the spatial nature of political relations and is necessary because, as Keil (2013, p. 796) argues in relation to the politics of urbanism, without accounting for the position of the urban as part of a wider set of relations, there is a danger of a "parochial localism" to this urban politics. It is not simply that authority is more multiple than an account of sovereignty may describe, but that it is expressed in more complex ways, forged through spatialities of folding and embeddedness. To develop a critical account of refugee urbanism, it is therefore not enough to note that cities are compositional arrangements of multiple authorities, in part because such an account of multiplicity is present in a range of work on governance, assemblages, and globalization (Deleuze & Guattari, 1987; Ferguson & Gupta, 2002; McFarlane, 2011). Rather, this requires both an examination of how those arrangements are embedded within, and enacted through, spatial relations, and a consideration of the varying intensities of governing and being governed that are brought to attention through urban conjunctures of governance, agency, and subjectivity.

The multiplicity of seeing like a city resonates with discussions over how authority is dispersed, rescaled, or "stretched" through new spatial formations and assemblages (Allen & Cochrane, 2010; Sassen, 2006). At the same time, such "unbundling" resonates with critical discussions of immigration enforcement as constituted through an array of contingent practices and authorities "constantly 'in formation' across a multiplicity of locations", rather than a coherent or coordinated model of state authority (Coleman, 2011, p. 309; Burridge et al., 2017; Mountz, 2010). However, Allen (2016, p. 21) argues that the concern with the rescaling of authority noted in much of this work, is often too blunt to account for how expressions of power have developed, as it remains tied to a topographic imaginary of spatial relations. In response, Allen (2016, p. 35) argues that exploring the critical possibilities of topology offers

> an opportunity to think again about what it actually means to say that events elsewhere are enfolded or woven into the political fabric of everyday life, or how it is that actors are able to make their presence felt in more or less powerful ways that cut across proximity and distance.

In seeing like a city, what "emerges is an understanding of the state as something that is perpetually in a position of becoming, whose institutional forms and capacities are always both transient and historically and geographically contingent" (Haughton et al., 2013,

p. 220). At the same time, there is a demand to look beyond the contingencies of the state to those other political actors, formations, and "arts of government" that are stabilizing temporary political orders. For Allen and Cochrane (2010, p. 1078), such orders are maintained through topological relations of influence and intensity that register power as a continuous, but transformed, presence in urban life. Presence in this sense may be physical or relational, as organizations may choose to be physically present but relationally distant, and in doing so may place themselves beyond reach as a means to displace or evade responsibility (Allen, 2016). Presence thus expresses the intensity of power relations, as influence is enacted over the conduct, decisions, and practices of other actors, agencies, and institutions. The city in this account is perceived as an arena in which diverse actors can make their presence felt, reflecting a "more *transverse* set of government interactions" (Allen & Cochrane, 2010, p. 1076, original emphasis), centered on the "ability to place others *within or beyond reach*" (Allen, 2016, p. 48, original emphasis).

In arguing for an approach to urban asylum orientated around refugee urbanism, I am arguing that we may better apprehend the complexities of how government is enacted by focusing on both the multiplicity of authorities *and* the relations of authority (and the forms they take). It is with this orientation that I now turn to consider the dispersal of asylum seekers in the UK.

## Dispersal and the city

Since 2000, the UK has operated a dispersal system for the accommodation of asylum seekers. After an initial period in temporary reception centers, asylum seekers are relocated to cities across the country to be accommodated whilst awaiting decisions on their asylum claims (Robinson et al., 2003). This system is almost exclusively an urban phenomenon, reflecting both the location of available low-cost housing and a presumption that for those successful in gaining refugee status, integration will be easier in more diverse urban centers. Yet work on the UK's dispersal system has failed to fully consider its urban character, preferring instead to focus on the regulatory basis of dispersal and its role at the heart of a policy of deterrence. In doing so, such discussions have overlooked how the conditions of urban life shape refugee experiences, and how cities and their politics are reworked not just by the presence of refugees, but by the presence of claims to authority, sovereignty, and governance that come with attempts to order mobility at different spatial scales. In the UK context, the regulatory role of dispersal means that the location and type of accommodation provided is on a "no choice" basis, and this lack of autonomy has been argued to mean that dispersal serves to further marginalize and socially isolate asylum seekers (Bloch & Schuster, 2005; Gill, 2009; Hynes, 2009). In this respect, the UK system bears similarities with dispersal policies in a number of other European countries (Robinson et al., 2003; Wren, 2003; Hamann, this special issue), with a key distinction being that in the UK dispersed asylum seekers are accommodated among urban communities rather than in segregated reception facilities. The UK might thus be argued to have resisted the urge, noted by Kreichauf (2018), for the "campization" of refugee infrastructure across Europe.

Through this more community-orientated model, accommodation for asylum seekers was, until 2012, provided either directly by local authorities, or through mixed consortiums that retained local authority involvement. In 2012 however, the Home Office signed a series of six centralized contracts passing accommodation and support to three private contractors – multinational companies G4S and Serco, and the accommodation partnership Clearel. These contracts were announced in the context of a government-wide austerity drive, and thus became a key means through which the then UK Border Agency sought to make "efficiency savings", whilst simultaneously removing local authority control over the process (Darling, 2016a, 2016b).

Dispersal has its origins in New Labour's investment in social infrastructure, and whilst never well funded, the policy did initially offer money to local authorities willing to work with the Home Office. It was first announced in the 1998 *Fairer, Faster, Firmer* White Paper, where it was described in the following terms;

> The administration of a new support scheme for asylum seekers . . . will require new national machinery to plan and co-ordinate provision, obtaining information from around the country and purchasing places either directly or by contracting with local agencies. Asylum seekers would be expected to take what was available, and would not be able to pick and choose where they were accommodated, but where possible placements would take account of the value of linking to existing communities and the support of voluntary and community groups . . . .This nationwide approach will help to relieve the burden on provision in London, where the majority of asylum seekers are currently concentrated (Home Office, 1998, p. 8.22).

Framed in this language of a "new national machinery" to accommodate asylum seekers, dispersal can be read as an imposition of new demands and expectations on local authorities and communities. It is this approach to examining dispersal that has orientated much prior work, focusing on dispersal as an expression of sovereign authority and the capacity of the state to regulate the lives of asylum seekers (see Bloch & Schuster, 2005; Darling, 2011; Hynes, 2009; Phillips, 2006). Dispersal in such work tends to be positioned as an expression of a politics of deterrence that centers on a desire to ensure the discomfort of those accommodated (for notable but localized exceptions, see Spicer, 2008; Wren, 2007). As such, critical attention has remained on the logics and expressions of this "machinery", viewing dispersal as a process that maintains the kind of classificatory work associated with seeing "like a state" as a means to render mobility legible and manageable (Scott, 1998). Such work is critically important, not least in highlighting the often repressive nature of dispersal as a technology of control. However, this presents only one perspective, and it is in this context, that I argue the insights of refugee urbanism may be productively mobilized.

The imaginary of dispersal as a policy practiced through a "new national machinery" of state and non-state actors working together to procure properties, provide support, and distribute populations, has a clear spatial dimension. For alongside the centralized imposition noted above, this "machinery" implies a power of distribution that is spatially "extensive" in Allen's (2016) terms. This is a power to assign positions and to place bodies, which extends outwards from the centralized bureaucracy of the Home Office in Croydon, and seeks to distribute asylum seekers geographically to avoid the "over-burdening" of London. Dispersal in this reading is partly an expression of sovereign authority, but also an inherently spatial and biopolitical matter of ordering populations (see Hannemann, this special issue). In Levine's (2015) terms, we might think of dispersal as an ordering principle or "form". For Levine, it is "the work of form to make order" (Levine, 2015, p. 3), and it is this that makes form political, as attention is drawn to a series of "ongoing contests over the proper places for bodies, goods, and capacities" (ibid; Rancière, 1999). Dispersal is an assertion of this "proper place" for specific types of bodies, yet in the imaginary of the "national machinery" the spaces in which bodies are distributed are undifferentiated, decontextualized, and largely absent. Dispersal becomes a process of extending power across the seemingly undifferentiated landscape of a "national approach". Whilst this may express the intended scope of dispersal as a distributive form, the realities of practicing such a form often involves instabilities as other logics come into contact with dispersal. It is these more contested foldings of power that are overlooked in focusing on the spatial extension of dispersal alone.

One starting point to address these contingencies is to consider dispersal as a political process that is productive in a number of ways. For, in establishing a system for accommodating asylum seekers, dispersal demanded the creation of new forms of knowledge, practices of support, and policy expertise among cities that had, until 2000, very little contact with forced migration. Prior to 2000, asylum seekers were able to seek support from local authorities of their choosing, leading many to rely on social networks in London. Thus, whilst asylum seekers were urban subjects in the UK before dispersal, their presence was often confined to a very limited range of urban settings. Whilst in theory, dispersal was established as a policy to ease a perceived "burden" on the south east of England, the reality has been one of significant regional and local disparities (Home Affairs Committee, 2018). In its earliest guises, this disparity came from a distribution of local authorities able to enter into agreements with the Home Office to accommodate asylum seekers. This required local authorities to have existing and available social housing stock that could be used at relatively short notice. The result was a landscape of dispersal focused on deindustrialized cities across the north of England, Scotland, and Wales. Cities like Glasgow, Leeds, Sheffield, and Newcastle viewed dispersal as a way to both meet a demand from the Home Office for accommodation, and enable otherwise hard-to-let social housing to generate an income.

Many of these cities were not accustomed to the support requirements of asylum seekers and were themselves suffering from pressures on social care and support services (Darling, 2016a, 2016b; Phillips, 2006). In this context, the "new national machinery" of dispersal led to the establishment of refugee community organizations across the UK, focused on supporting the needs of newly arrived asylum seekers (Griffiths et al., 2006). At the same time, dispersal in practice rapidly became entwined with a parallel process of reporting, through which asylum seekers were required to regularly register their presence with an immigration reporting center or, more often, with a local police station (Burridge & Gill, 2017). This regulatory mechanism has served to not only restrict the mobility of asylum seekers, but together with the variable landscape of legal support across dispersal cities, has exposed asylum seekers to greater risk of detention and deportation (Fisher et al., 2019). If we consider dispersal as a form in Levine's (2015, p. 3) terms, as an "arrangement of elements" that orders, shapes, and distributes lives and social worth, then part of this process was the construction of an architecture of support and reporting at the urban level to maintain and reproduce this very system of ordering.

Considering dispersal as a productive "art of government" that has created the conditions for new forms of knowledge and expertise has value. However, this gesture still remains entangled with a sovereign view of dispersal as a disciplinary form that is territorial in its orientation, and spatially extensive in its expression of authority. To move beyond this, there is a need to consider how dispersal alters as its conditions of ordering and distribution fluctuate.

## Negotiating dispersal

To see dispersal "like a city" is to focus on the multiple claims to authority and influence that may be argued to shape dispersal, and to consider how authority is embedded in relations that are not simply "extensive" in spatial terms (Allen, 2016). A more uncertain and complex narrative on dispersal offers insights that a concern with dispersal as a disciplinary form obscures. To make this case, I want to consider an example drawn from Glasgow, building on the experiences of Anna with which I opened this paper.

At the end of July 2018, Glasgow became the focal point for media scrutiny of the UK's treatment of asylum seekers and refugees.[1] The attention centered on an announcement from the international services firm Serco that they would be restarting a controversial programme of lock-changes on properties used to accommodate asylum seekers in the city, as a means of forcibly evicting those who had reached the end of the asylum process. Despite Serco arguing that only those asylum seekers who had their claims rejected would be evicted, asylum advocates asserted that the poor decision-making record of the Home Office would inevitably mean that individuals qualifying for refugee status would be made homeless (see Shaw & Kaye, 2013). The subsequent weeks of public debate, demonstration, and dissent drew attention to two prominent factors in UK asylum policy – the outsourcing of housing away from urban authorities, and the production of destitution among asylum seekers in the hope of forcing them to leave the country.

In the wake of Serco's announcement, a number of responses emerged, illustrating the role of multiple political actors seeking to negotiate the position of asylum seekers in the city, albeit for potentially contrasting reasons. First, the local authority appealed to the Home Office, warning of a homelessness crisis on its streets if the policy was followed. Whilst politically significant, this intervention was nevertheless limited as the council is legally prohibited from housing asylum seekers (Brooks, 2018). Second, a range of refugee rights' groups, legal advocates, and campaigners, staged public demonstrations in the city, burning eviction notices and asserting the right to remain. These demonstrations raised public awareness of the case, but also called for action from the council and housing providers to step in to avoid a homelessness crisis. Third, a housing association sub-contracted by Serco responded after public pressure, announcing that they would refuse any request from Serco to change locks. In doing so, their director of operations stated that they had;

> written to Serco advising that as per the terms of the lease agreement they require our written permission to alter fixtures and fittings, and advised Glasgow city council that we will rehouse our local Serco tenants if evicted (cited in Brooks, 2018).

In doing so, this housing association was mobilizing what was locally known as the "flipping model", whereby a Serco occupancy agreement is converted into a temporary social housing tenancy without the occupant having to relocate (Brooks, 2018). Flipping was a discretionary practice far more common under previous models of public provision of accommodation. Local authorities could convert a property from dispersal accommodation to social housing at the point at which an individual received refugee status or required emergency accommodation following refusal. From the local authority perspective, the value of "flipping" was in allowing an individual to remain in a property without relocation or homelessness, thus allowing time for longer-term solutions to be found. Since 2012, however, such discretion over housing has largely been lost as Serco and other private providers have no incentive to work with local authorities in such a way, and see the practice as a loss of potential income (Darling, 2016a).

Following this period of pressure, Serco were forced to pause the decision to change locks and are, at the time of writing, seeking legal advice alongside the Home Office, as to how to proceed. This is a moment of reprieve rather than a victory for those supporting the rights of asylum seekers in the city. Nevertheless, it draws attention to "how all of the apparent seamless accomplishment of urban change are full of messy encounters, deals, and protracted and often failed negotiations, and that high-speed urban remaking is often full of holes to be occupied and shaped" (Simone & Pieterse, 2017, p. 181). It is the messiness of this urban politics, its holes and failed negotiations, that is overlooked by a concern with dispersal as a disciplinary form. With this in mind, I want to consider three ways in which "seeing like a city" might illustrate different dimensions of the Glasgow case, drawing on the three implications of "seeing like a city" discussed previously.

### *Claims to authority*

First, we might see a range of competing claims to authority in effect in Glasgow. Here Serco enters into a new domain of operations – asylum housing – as an actor without prior experience. Whilst supported by the Home Office, it quickly became clear that Serco were not able to pursue their policy of lock-changing without the assent of other actors. As Allen (2016) argues, for authority to be effective it has to be recognized and responded to by others. The authority of Serco relied on the adherence of a range of subcontractors tasked with putting lock-changes into practice. As a result, Serco's ability to push through such a policy was questioned, as other actors began to assert their own claims to authority, including the local authority's appeal to the Home Office, the refusal of housing associations to meet the terms of their contract, and the mobilization of public protests and media coverage of the evictions (BBC, 2018). The halt in lock-changes that arose might thus be read as an expression of an urban politics that is necessarily negotiated.

It is precisely this negotiated character that Magnusson (2011) and Valverde (2011) draw to the fore, arguing that whilst there is often a pragmatism to urban politics, there is also a restlessness as issues are rarely "settled", but rather present temporary compromises (Simone & Pieterse, 2017). The discretionary practice of "flipping" properties in place in Glasgow prior to 2012, was one such temporary stabilization. It did not address the challenges of refugee homelessness or the use of destitution as a policy tool directly, but offered a negotiated means of temporarily mitigating these effects. Addressing homelessness was a deferred issue. It is this deferral that an approach to seeing like a city brings to the fore, to argue that in a context of multiple authorities, deferral and the temporary stabilization of a set of rights is inevitable. To focus on multiple authorities is not to suggest that such authorities are equal or to dismiss the hierarchies of power through which dispersal operates. Rather, the point is to reflect on how those assemblages of authority, rule, and dissent, operate in and through the city, and how they exceed an image of dispersal as a disciplinary form that extends "into" a city like Glasgow.

### *Intensities of authority*

The second insight to draw out, is that whilst seeing like a city may focus on how multiple authorities shape dispersal in practice, these are often spatially diffuse actors and rely upon a range of spatial tactics to produce their authority. The Glasgow case thus illustrates the topological maintenance of relations of authority as a politics of "reach" enables physically distant authorities to shape negotiations in the city. In this context;

> Reach … is a form of relational distance – a space composed by the stretching, folding and distorting of relationships to achieve certain ends. As such, reach can be leveraged topologically in a variety of distinctive ways by cutting across proximity and distance to register a powerful presence of one particular kind or another. The quieter, more muted registers of power, increasingly privileged by institutions and social groups alike, arguably owe much to such leverage. (Allen, 2016, p. 49)

These topological relations of "reach" were reshaped with the privatization of asylum housing in two senses. First, in the ways in which different actors are brought into relational proximity despite physical distance. The outsourcing of dispersal to a series of private security contractors and their sub-contractors offers a means through which direct governmental authority may be physically absent from the negotiations of dispersal in urban life in Glasgow. Yet despite this, government still reaches into such negotiations to shape and direct them. This describes a process whereby "government initiatives are 'detached' from the centre and 'lodged' in subnational institutions and agencies to secure compliance", resulting in authority being "not so much consented to as embedded" in new actors (Allen, 2016, p. 50). The effect, Allen (ibid) suggests, is to "stabilise a pattern of authority at arm's length, by aligning local rulings and directives with those of the 'centre'", and forging a mediated "immediacy of presence" for those "who by and large are detached". In Glasgow, whilst the physical presence of Home Office officials, or even representatives of Serco themselves, may have been absent, these actors were "co-present in [the] power arrangement taking shape" (ibid, p. 43), as they directed negotiations, shaped contractual obligations, and influenced discussions through mediated relations with sub-contractors, local authorities, and local media. In this way, government is able to bring the issue of asylum dispersal in Glasgow into relational "reach" whilst remaining physically distant from its effects in the city. As a topological reading implies, we see a process in which the expression of power is able to "change yet nonetheless remain the same" (Allen, 2016, p. 13), as despite the multiplicity of claims to authority evident in Glasgow, government still retains a significant, but not all-encompassing, capacity to shape negotiations.

The second effect of outsourcing is in this relation of retained authority but increasing distance. The very practice of outsourcing represents not just a transfer of responsibility and accountability from government to private sector contractors, but also a process of "lengthening" the pathways of mediation between an issue and the government (Darling, 2016a). Thus, a politics of "reach" may also act to place political concerns "out of reach" through "*stretching out* the exchanges between actors" (Allen,

2016, p. 54, original emphasis). The mediating role of private forms of authority is evident in the manner in which Serco and their sub-contractors became the focus of political and public anger, rather than this being directed at the decision-making of the Home Office and the use of destitution as a policy tool in the first instance. As with a number of other scandals involving the conditions of asylum accommodation in the UK, the Government is able to remove their proximity from questions of accountability by creating a relational distance. By investing private actors with authority as a means to "reach" into dispersal locations while remaining physically distant, the state is able to displace their own authority from such contexts and displace their responsibilities for asylum seekers. A focus on seeing like a city is fundamental to drawing out this uneven landscape of *displacing and emplacing* authority by considering how authority "emerges" in and through different urban contexts and compositions (Magnusson, 2011, p. 36).

### *Unintentional effects*

Seeing like a city may also focus attention on the often unintentional effects of dispersal, and how these present opportunities for dissent within the topological workings of power. For example, in the Glasgow case, dispersal has produced not just the development of the skills and expertise noted earlier, but also a more assertive politics of dissent that demands rights for asylum seekers. Thus, in 2009, activists in Glasgow and Bristol founded the "Dignity not Destitution" campaign, aimed at challenging government policy on the removal of support for asylum seekers after decisions on their status have been taken. In opposing the use of destitution as a deterrent within the asylum system, the "Dignity not Destitution" campaign argued that asylum seekers should be provided with sufficient support so that they can meet their essential living needs and be given permission to work if their case hasn't been resolved within six months. The campaign was focused on addressing demands to national government *through the city* as a conduit for advancing a refugee rights agenda. Through public demonstrations, lobbying of councilors, and public petitions, the campaign focused on galvanizing support within urban authorities and mobilizing this against "central" government (Darling, 2016). One unintended consequence of dispersal is therefore that dispersal produced the conditions within which such opposition to government policy could be fostered, and alliances of interest between activists, local authorities, and asylum seekers could be forged. In the latter instance, campaigns and dissent emerged in large part through the collaboration of citizen and non-citizen activists, with a lens of solidarity focusing on points of shared exclusion, most notably the ways in which state-led austerity enacts violence across the city (Darling, 2016b). From efforts by asylum seekers to monitor and record their inadequate and unsafe housing conditions, to anti-deportation activism in support of fellow asylum seekers, forms of "minor" politics of registering dissent and seeking recognition of political presence, were often the starting points on which such campaigns were built (Darling, 2017; Squire and Darling 2013).

Dispersal also produced cultural effects that exceed the form of a disciplinary practice. In 2015, the charity "Refuweegee" was established in Glasgow aiming to provide a welcome to asylum seekers and refugees through a series of volunteering opportunities, the distribution of donated household items, and the provision of welcome packs to all new arrivals in the city (Refugweegee, 2017). A "refuweegee" is defined as "a person who upon arrival in Glasgow is embraced by the people of the city, a person considered to be local", under the banner "We're all fae somewhere", communicating a message that what matters is not the process that placed asylum seekers and refugees in Glasgow, but their presence in the city here and now (Refugweegee, 2017). Dispersal created not just conditions for dissent, but also conditions for the emergence of fragile and often temporary cultures of welcome in everyday urban life (Darling, 2018; Gill, 2018). Seeing like a city allows for these uncertain effects of dispersal to come to the fore, wherein the cultural responses of Glasgow, together with its political histories of solidarity and internationalism (Featherstone, 2012), combine with the lived realities of dispersal to present opportunities for challenging dominant narratives of asylum as an issue of securitization, regulation, and suspicion. Taken together, these two cultural and political claims speak to the importance of political "experiments at the level of everyday life", as Simone and Pieterse (2017, p. 9) assert. For them, urban politics is not about the establishment of an all-encompassing framework of dissent or political antagonism, but rather of more prosaically exploring how "topological resonances" between actors, ideas, issues, and materials, may "open up the possibility of new linkages among things usually kept apart" (ibid, p. 10). These resonances are the relations of intensity, reach, and influence that we see topologically reproducing power in the assemblages of dispersal, yet they also offer folds of activism, solidarity, and alliance, between political causes and campaigns, such that the issue of destitution in Glasgow becomes a point of learning for asylum advocates elsewhere in the UK. In this way, the urban intra-actions of powerful relations are made vulnerable to deformation and appropriation in new directions.

## Conclusion

In this paper, I have explored how urban asylum might be approached by employing an emergent set of debates on what it may mean to see like a city. This approach foregrounds the incomplete nature of claims to authority, and resonates with a view of the nation-state as an assemblage "that is perpetually in a position of becoming, whose institutional forms and capacities are always both transient and historically and geographically contingent" (Haughton et al., 2013, p. 220). More than this, seeing like a city highlights the political possibilities of the urban to challenge the remaking of sovereign authority, and views diverse patterns of everyday urban life as a resource from which to build such possibilities. To more fully account for refugee urbanism though, I have argued that seeing like a city as a political ontology should be supplemented with a critical appreciation of the relations of intensity, reach, and influence that shape how conjunctures of urban governance, agency, and subjectivity co-constitute "urban refugees". Critically, as the context of dispersal illustrates, the possibilities that seeing like a city highlights in its account of the "navigable channels" of responding to, and resisting, governance are not equally or evenly spread. Rather, the capacity to respond to the governmental impulses of multiple authorities and select one's own path through these, is a privileged position. What seeing like a city may help to expose, is the ways in which decisions over that inequality are made not by a singular authority, but by the interaction of a multitude of sources of government, influence, and authority. Such multiplicities involve the folding in and stretching out of power relations and claims, such that cities are spaces of unpredictable effects and politics.

There are of course limits to "seeing like a city", not least in the risk of romanticizing how authority may be produced through collective claims-making and prospective self-government (Magnusson, 2011, 2015). Nor should this be read as an argument that dismisses state sovereignty or authority as of real significance in refugee politics (Darling, 2017). Indeed, focusing on dispersal presents a critical challenge to the concern with transformation and instability that seeing like a city tends to foreground. For whilst in this political ontology orders may be seen as temporary stabilizations in aleatory conditions (Dunn & Cons, 2014), it is important to recognize that seeking stability within conditions of flux is not simply a political product of sovereign, or governmental authority. A range of urban actors seek stability too. In the case of Anna, alongside her demands for dignity, safety and basic information on her situation, there is a desire for stillness and stability in the turbulent conditions of being made mobile by dispersal. Anna invested considerable amounts of her labor and sense of self into creating some stability for herself and her family, speaking to the risks that arise through urban dynamics of instability and aleatory authority. As Dunn and Cons (2014) note, whilst the complexity of overlapping forms of governance may offer openings for evasion and unintended sites of resistance, such effects are rarely without cost. Dispersal thus highlights how the hopeful politics that Magnusson (2011) attaches to the city as a space of transformation, is challenged by the disruptive nature of that transformation at the level of citizens and non-citizens who experience the politics of the state as intentionally disruptive and discomforting (Darling, 2011). This calls for a nuanced approach to seeing like a city, which is not content to rely on the potentials of urban multiplicity and uncertainty to unproblematically undermine the exclusions of state sovereignty and bordering. Rather, in considering dispersal, the need to retain a concern with how the capacity to navigate tensions of uncertainty and stability is unevenly distributed is one means of developing an account of refugee urbanism that eschews urban romanticism or exceptionalism.

In developing such an account, and despite its challenges, seeing like a city does present a number of opportunities. First, methodologically, a seeing like a city approach entails a concern for staying with multiplicity and focusing attention on the temporary stabilization of orders. In the context of urban asylum, this means examining not just the disciplinary mechanisms and technologies that shape and condition how asylum seekers encounter cities, but also the spatial and social practices through which urban experience is constituted. Addressing multiplicity through such practices demands renewed attention to the materials and infrastructures that shape how refugees and asylum seekers experience cities, from governmental materials of categorization (Darling, 2014), through infrastructures of accommodation and urban services (Amin & Thrift, 2017), to material and infrastructural failures that express relations of power and position within the city, such as the collapsed ceiling and its aftermath for Anna (Lancione & McFarlane, 2016). It also demands greater critical engagement with how urban worlds are learned and adapted to as complex compositions of authority, autonomy, presence, and aversion. The multiplicity that seeing like a city foregrounds is an ontology of uncertainty and, as such, requires attention to how uncertainty is lived with and adapted to as a mode of cohabiting in the midst of urban "throwntogetherness" (Massey, 2005).

Secondly, seeing like a city may offer a way of approaching political questions, and not just a way of understanding the urban. This is important as it shifts the kind of political claims and demands that are made – from either abandoning the state altogether as a political formation or to radically altering its constitution through political revolution, to an account of situated practices and counter-conducts that emerge *within* existing fields of government. As the forms of dissent that emerged in Glasgow illustrate, these practices are often unpredictable effects of generative forms of power and claims to authority. Such an approach is evident in Larner's (2014, p. 197) discussion of whether the mechanisms of creative appropriation often used by the state and other dominant actors to side-line and incorporate dissent, might also be employed by those seeking to challenge such dominance, thereby asking if it is possible to "imagine new 'arts of government' that might take advantage of (instead of simply denouncing or resisting) recent transformations in the spatial organisation of government" (ibid). Refugee urbanism may offer one perspective from which to start imagining these possibilities in the context of urban asylum.

## Note

1. For example, coverage in a range of newspapers drew attention to Glasgow as a city at the forefront of negotiating the UK's asylum system. These included headlines of "Being a 'failed asylum seeker' leads directly to homelessness on the streets of Glasgow" (Clark, *The New Statesman*, 2018), "Confusion over 'eviction' of Glasgow's asylum seekers" (Goodwin, *The Herald*, 2018), and 'Calls to halt plans to change locks on homes of 300 Glasgow asylum seekers as city could plunge into "humanitarian crisis" (Paterson, *The Scottish Sun*, 2018).

## Acknowledgements

Thanks to all those who gave their time and insight to the research on which this paper is based. An earlier version of this paper was presented at the 'Infrastructures of Arrival' workshop hosted by Indiana University Bloomington. My thanks to Elizabeth Cullen Dunn for the invitation and to Romola Sanyal for her extensive commentary on this paper. Thanks to three anonymous reviewers for their constructive feedback, and to Rene Kreichauf and Birgit Glorius for their comments as editors of this special issue. Finally, the paper has benefited greatly from the intellectual support and insight of a number of colleagues and friends, namely Louise Amoore, Ben Anderson, Gordon MacLeod, Colin McFarlane and Helen Wilson. All errors remain my own.

## Disclosure statement

No potential conflict of interest was reported by the author.

## Funding

This work was supported by the Economic and Social Research Council [ES/K001612/1].

## References

Allen, John. (2016). *Topologies of power: Beyond territory and networks.* London: Routledge.

Allen, John, & Cochrane, Allan. (2010). Assemblages of state power: Topological shifts in the organization of government and politics. *Antipode*, *42*(5), 1071–1089. https://doi.org/10.1111/j.1467-8330.2010.00794.x
Amin, Ash, & Thrift, Nigel. (2017). *Seeing like a city*. Cambridge: Polity.
Bagelman, Jennifer. (2016). *Sanctuary city: A suspended state*. Basingstoke: Palgrave Macmillan.
Bauder, Harald. (2017). Sanctuary cities: Policies and practices in international perspective. *International Migration*, *55*(2), 174–187. https://doi.org/10.1111/imig.12308
BBC (2018). *Home Secretary urged to halt eviction of failed asylum seekers in Glasgow. BBC News*. Retrieved October 10, 2019, from. https://www.bbc.co.uk/news/uk-scotland-glasgow-west-45007011
Bloch, Alice, & Schuster Liza. (2005). At the extremes of exclusion: Deportation, detention and dispersal. Ethnic and Racial Studies, *28*(3), 491–512. https://doi.org/10.1080/0141987042000337858
Brooks, Libby. (2018). Serco to pause plans to evict asylum seekers after glasgow protests. The Guardian, 4 August 2018, (last accessed 15 June 2020). Available at: https://www.theguardian.com/uk-news/2018/aug/04/glasgow-protesters-rally-against-serco-plan-to-evict-asylum-seekers
Burridge, Andrew, & Gill, Nick. (2017). Conveyor-belt justice: Precarity, access to justice, and uneven geographies of legal aid in UK asylum appeals. *Antipode*, *49*(1), 23–42. https://doi.org/10.1111/anti.12258
Burridge, Andrew, Gill, Nick, Kocher, Austin, & Martin, Lauren. (2017). Polymorphic borders. *Territory, Politics, Governance*, *5*(3), 239–251. https://doi.org/10.1080/21622671.2017.1297253
Coleman, Matthew. (2011). Topologies of practice. *Dialogues in Human Geography*, *1*(3), 308–311. https://doi.org/10.1177%2F2043820611421549 3 doi:10.1177/2043820611421549
Darling, Jonathan. (2011). Domopolitics, governmentality and the regulation of asylum accommodation. *Political Geography*, *30*(5), 263-271. https://doi.org/10.1016/j. *Political Geography*, *2011 2011 30*(04 5), 011. doi:10.1016/j.polgeo.2011.04.011
Darling, Jonathan. (2014). Another letter from the Home Office: reading the material politics of asylum. *Environment and Planning D: Society and Space*, *32*, 484–500. https://doi.org/10.1068%2Fd13067p
Darling, Jonathan. (2016). Defying the demand to 'go home': from human rights cities to the urbanization of human rights. In Barbara Oomen, Martha. F. Davis & Michele Grigolo (Eds.)., (pp. 121–38). *Global urban justice: The rise of human rights cities*. Cambridge: Cambridge University Press
Darling, Jonathan. (2016a). Privatising asylum: Neoliberalisation, depoliticisation and the governance of forced migration. *Transactions of the Institute of British Geographers*, 41(3), 230-243. https://doi.org/10.1111/tran.12118
Darling, Jonathan. (2016b). Asylum in austere times: Instability, privatization and experimentation within the UK asylum dispersal system. *Journal of Refugee Studies*, *29*(4), 483–505. https://doi.org/10.1093/jrs/few038
Darling, Jonathan. (2017). Forced migration and the city: Irregularity, informality, and the politics of presence. *Progress in Human Geography*, *41*(2), 178–198. https://doi.org/10.1177%2F0309132516629004
Darling, Jonathan. (2018). The fragility of welcome – commentary to Gill. *Fennia*, *196*(2), 220–224. https://doi.org/10.11143/fennia.75756 2 doi:10.11143/fennia.75756
Darling, Jonathan, & Bauder Harald. (2019). (Eds.) *Sanctuaries cities and urban struggles: Rescaling migration, citizenship, and rights*. Manchester: Manchester University Press.
Darling, Jonathan, & Wilson, Helen, F. (2016). (Eds.) *Encountering the City: Urban Encounters from Accra to New York*. London: Routledge.
Deleuze, Gilles, & Guattari, Felix. (1987). *A thousand plateaus: Capitalism and schizophrenia*. Translated by Brian Massumi. Minneapolis: University of Minnesota Press.
Depraetere, Anika, & Oosterlynck, Stijn. (2017). "I finally found my place": A political ethnography of the Maximiliaan refugee camp in Brussels. *Citizenship Studies*, *21*(6), 693–709. https://doi.org/10.1080/13621025.2017.1341653

Dunn, Elizabeth Cullen. (2012). The chaos of humanitarian aid: Adhocracy in the Republic of Georgia. *Humanity*, *3*(1), 1–23. https://doi.org/10.1353/hum.2012.0005

Dunn, Elizabeth Cullen, & Cons, Jason. (2014). Aleatory sovereignty and the rule of sensitive spaces. *Antipode*, *46*(1), 92–109. https://doi.org/10.1111/anti.12028

Featherstone, David. (2012). *Solidarity: Hidden histories and geographies of internationalism*. London: Zed Books.

Ferguson, James, & Gupta, Akhil. (2002). Spatializing states: Toward an ethnography of neoliberal governmentality. *American Ethnologist*, *29*(4), 981–1002. https://doi.org/10.1525/ae.2002.29.4.981

Fisher, Daniel X.O., Burridge, Andrew, & Gill, Nick. (2019). The political mobilities of reporting: Tethering, slickness and asylum control. *Mobilities*, *14*(5), 632–647. https://doi.org/10.1080/17450101.2019.1607049

Foucault, M. (2003). *Society Must be Defended: Lectures at the College de France*, 1975–1976. London: Picador

Gill, Nick. (2009). Governmental mobility: the power effects of the movement of detained asylum seekers around Britain's detention estate. *Political Geography*, *2009 2009 28*(05 3), 003. doi:10.1016/j.polgeo.2009.05.003

Gill, Nick. (2010). New state-theoretic approaches to asylum and refugee geographies. *Progress in Human Geography*, *34*(5), 626–645. https://doi.org/10.1177/0309132509354629

Gill, Nick. (2018). The suppression of welcome. *Fennia*, *196*(1), 88–98. https://doi.org/10.11143/fennia.70040

Griffiths, David, Sigona, Nando, & Zetter, Roger. (2006). Integrative paradigms, marginal reality: Refugee community organisations and dispersal in Britain. *Journal of Ethnic and Migration Studies*, *32*(5), 881–898. https://doi.org/10.1080/13691830600704529

Haughton, Graham, Allmendinger, Phil, & Oosterlynck, Stijn. (2013). Spaces of neoliberal experimentation: Soft spaces, postpolitics, and neoliberal governmentality. *Environment & Planning A*, *45*(1), 217–234. https://doi.org/10.1068/a45121

Home Affairs Committee (2018). *Oral evidence: Asylum accommodation*. Retrieved from. http://data.parliament.uk/writtenevidence/committeeevidence.svc/evidencedocument/home-affairs-committee/asylum-accommodation/oral/92748.html

Home, Office. (1998). *Fairer, faster and firmer – a modern approach to immigration and asylum*. London: Home Office. Available at https://www.gov.uk/government/publications/fairer-faster-and-firmer-a-modern-approach-to-immigration-and-asylum (last accessed 15 June 2020).

Hynes, Patricia. (2009). Contemporary compulsory dispersal and the absence of space for the restoration of trust. *Journal of Refugee Studies*, *22*(1), 97–121. https://doi.org/10.1093/jrs/fen049

Keil, Roger. (2013). 'At one remove'? Warren Magnusson sees like a city and finds the world. *International Journal of Urban and Regional Research*, *37*(2), 793–797. https://doi.org/10.1111/1468-2427.12014

Koptyaeva, Alexandra. (2017). Collective homemaking in transit. *Forced Migration Review*, *55*, 37–38.(last accessed 15 June 2020). https://search.proquest.com/docview/1929035655/fulltext/ABDAEA584DD44724PQ/1?accountid=14533

Kreichauf, René. (2018). From forced migration to forced arrival: The campization of refugee accommodation in European cities. *Comparative Migration Studies*, *6*(7). https://doi.org/10.1186/s40878-017-0069-8

Lancione, Michele, & McFarlane, Colin. (2016). Infrastructural becoming: sanitation, cosmopolitics, and the (un)making of urban life at the margins. In Anders Blok & Ignacio Farías (Eds), *Urban cosmopolitics: agencements, atmospheres, assemblies* (pp. 45-62). London: Routledge.

Lancione, Michele. (2016). Racialised dissatisfaction: Homeless management and the everyday assemblage of difference. *Transactions of the Institute of British Geographers*, *41*(2), 363–375. https://doi.org/10.1111/tran.12133

Larner, Wendy. (2014). The limits of post-politics: Rethinking radical social enterprise. In Japhy Wilson & Erik Swyngedouw (Eds.), *The post-political and its discontents: Spaces of depoliticisation, spectres of radical politics* (pp. 189–207). Edinburgh: Edinburgh University Press.

Lefebvre, Henri. (2003). *The urban revolution.* (Translated by Robert Bononno). Minneapolis: University of Minnesota Press.
Levine, Caroline. (2015). *Forms: Whole, rhythm, hierarchy, network.* Princeton: Princeton University Press.
Maestri, Gaja. (2017). Struggles and ambiguities over political subjectivities in the camp. *Citizenship Studies, 21*(6), 640–656. https://doi.org/10.1080/13621025.2017.1341656
Magnusson, Warren. (2011). *Politics of urbanism: Seeing like a city.* London: Routledge.
Magnusson, Warren. (2014). The symbiosis of the urban and the political. *International Journal of Urban and Regional Research, 38*(5), 1561–1575. https://doi.org/10.1111/1468-2427.12144
Magnusson, Warren. (2015). *Local self-government and the right to the city.* Montreal: McGill-Queens University Press.
Massey, Doreen. (2005). *For space.* London: Sage.
Massey, Doreen. (2007). *World city.* Cambridge: Polity.
McFarlane, Colin. (2011). The city as assemblage: Dwelling and urban space. *Environment and Planning D, Society & Space, 29*(4), 649–671. https://doi.org/10.1068/d4710
Mountz, Alison. (2010). *Seeking Asylum: Human smuggling and bureaucracy at the border.* Minneapolis: University of Minnesota Press.
National Audit Office. (2014). COMPASS contracts for the provision of accommodation for asylum seekers House of Commons, available at: https://www.nao.org.uk/wp-content/uploads/2014/01/10287-001-accommodation-for-asylum-seekers-Book.pdf (last accessed 15 June 2020)
Nordling, Vanna, Sager, Maja, & Söderman, Emma. (2017). From citizenship to mobile commons: Reflections on the local struggles of undocumented migrants in the city of Malmö, Sweden. *Citizenship Studies, 21*(6), 710–726. https://doi.org/10.1080/13621025.2017.1341660
Phillips, Deborah. (2006). Moving towards integration: The housing of asylum seekers and refugees in Britain. *Housing Studies, 21*(4), 539–553. https://doi.org/10.1080/02673030600709074
Picker, Giovanni. (2017). *Racial cities: Governance and the segregation of Romani people in urban Europe.* London: Routledge.
Rancière, Jacques. (1999). *Dis-agreement: Politics and philosophy* (translated by Julie Rose). Minneapolis: University of Minnesota Press
Refugweegee. (2017) *Refuweegee website.* Retrieved October 28, 2018, from. https://www.refuweegee.co.uk/
Robinson, Vaughan, Andersson, Roger, & Musterd, Sako. (2003). *Spreading the 'burden'? A review of policies to disperse asylum seekers and refugees.* Bristol: Policy Press.
Rodgers, Scott, Barnett, Clive, & Cochrane, Allan. (2014). Where is urban politics? *International Journal of Urban and Regional Research, 38*(5), 1551–1560. https://doi.org/10.1111/1468-2427.12143
Sanyal, Romola. (2014). Urbanizing refuge: Interrogating spaces of displacement. *International Journal of Urban and Regional Research, 38*(2), 558–572. https://doi.org/10.1111/1468-2427.12020
Sanyal, Romola. (2018). Managing through ad hoc measures: Syrian refugees and the politics of waiting in Lebanon. *Political Geography, 66*(1), 67–75. https://doi.org/10.1016/j.polgeo.2018.08.015
Sassen, Saskia. (2006). *Territory, authority, rights: From medieval to global assemblages.* Princeton: Princeton University Press.
Scott, James, C. (1998). *Seeing like a state.* New Haven: Yale University Press.
Scottish Refugee Council. (2014). The extent and impact of asylum accommodation problems in Scotland available at: https://www.scottishrefugeecouncil.org.uk/wp-content/uploads/2019/10/The-Extent-and-Impact-of-Asylum-Accommodation-Problems-in-Scotland-full-report-August-2014.pdf (last accessed 15 June 2020)
Shaw, Jan., & Kaye, Mike. (2013). *A question of credibility: Why so many initial asylum decisions are overturned on appeal in the UK.* London: Amnesty International
Simone, Abdou Maliq, & Pieterse, Edger. (2017). *New urban worlds: Inhabiting dissonant times.* Cambridge: Polity Press.

Spicer, Neil. (2008). Places of exclusion and inclusion: Asylum-seeker and refugee experiences of neighbourhoods in the UK. *Journal of Ethnic and Migration Studies, 34*(3), 491–510. https://doi.org/10.1080/13691830701880350

Squire, Vicki., & Darling, Jonathan. (2013). The “minor” politics of rightful presence: Justice and relationality in City of Sanctuary. *International Political Sociology, 7*(1), 59–74. https://doi.org/10.1111/ips.12009

Taylor, Zack. (2013). Magnusson’s challenge: Can political science learn to ‘see like a city’? *International Journal of Urban and Regional Research, 37*(2), 800–802. https://doi.org/10.1111/1468-2427.12014

Valverde, Mariana. (2011). Seeing like a city: The dialectic of modern and premodern ways of seeing in urban governance. *Law & Society Review, 45*(2), 277–312. https://doi.org/10.1111/j.1540-5893.2011.00441.x

Wilson, Helen, F. (2017). On geography and encounter: Bodies, borders, and difference. *Progress in Human Geography, 41*(4), 451–471. https://doi.org/10.1177/0309132516645958

Wirth, Louis. (1938). Urbanism as a way of life. *American Journal of Sociology, 44*(1), 1–24. https://doi.org/10.1086/217913

Wren, Karen. (2003). Refugee dispersal in Denmark: From macro- to micro-scale analysis. *International Journal of Population Geography, 9*(1), 57–75. https://doi.org/10.1002/ijpg.273

Wren, Karen. (2007). Supporting asylum seekers and refugees in Glasgow: The role of multi-agency networks. *Journal of Refugee Studies, 20*(3), 391–413. https://doi.org/10.1093/jrs/fem006

Zeiderman, Austin, Kaker, Sabia, Silver, Jonathan, & Wood, Astrid. (2015). Uncertainty and urban life. *Public Culture, 27*(2), 281–304. https://doi.org/10.1215/08992363-2841868

# Towards a parallel exceptional welfare system: the scaling down and out of forced migrants' reception in Italy

Michela Semprebon

**ABSTRACT**
Reception measures have been implemented in many EU countries, as part of an articulated framework of European, national, subnational, local normative arrangements. This paper states the need for more attention on the division of responsibility among government levels and forms of contention associated with it, moving beyond the Scholten's framework (2013) focused on actors' coordination. In doing so, it argues that an increased focus on the welfare perspective can illuminate how a parallel, exceptional system of welfare has been created and consolidated*ad hoc* for forced migrants, with potentially negative impacts on their social protection. The argument builds on empirical evidence collected in Italy in the period 2014-2020. Beyond the specificity of the Italian case, the paper aims to put forward a more general discussion on how the literature on welfare can provide a useful theoretical background to analyse the relationship between forced migrants and the city.

## Introduction

This paper focuses on reception schemes that are an integral part of this policy scenario in many EU Member States (for Italy, see Marchetti 2014; Balbo 2015; Campomori & Ambrosini, 2020; Semprebon and Pelacani 2020; for an overview on EU countries more generally: Glorius and Doormenik, 2016; Glorius and Doormenik, 2020). Such schemes comprise an articulated policy framework that involves several government agencies at different territorial levels, with a prominent role of the local level. While the configurations theorized by Scholten (2013), and the subsequent debate, have been useful in detailing the functioning of the governance of migration, less attention has been devoted to the actual division of responsibility among agencies, the contentious dynamics (Campomori & Ambrosini, 2020) that have characterized it and the potential impact on forced migrants – who for the scope of this paper include specifically holders and seekers of international protection – particularly in the long-term.

Analyses of the division of responsibility across social policies have highlighted processes of rescaling in many European member states (Ferrera, 2005; Kazepov, 2010). This paper argues that such processes provide a useful theoretical framework to explore similar patterns of downward rescaling in reception policies and to show how an exceptional welfare system has been created ad hoc for forced migrants and has been strengthened throughout the second decade of the new millenium. The exceptional "emergency" nature of many of these polices has been often underlined particularly with reference to increasing flows from Lybia in 2011 (Campesi & The Arab Spring and the Crisis of the European Border Regime: Manufacturing Emergency in the Lampedusa Crisis, 2011; Dal Zotto, 2014; Marchetti, 2012), to national and local policies more generally (Scotto, 2018; Pitzalis, 2018; Della Puppa et al., 2020) and to reception structures (Vianelli, 2014). The contentious nature of these governance dynamics, and its strict connection with welfare needs and claims, has been less explored. Yet it throws further light on the risks of inadequate social protection for forced migrants, particularly in the post-reception phase. In this direction, this paper will feed the academic debate on reception policies, by reflecting on their interconnection with welfare rights, that have long been associated with the residency status in the polis.

The paper is structured as follows. First, it will briefly review the literature on migration and the role of local authorities in the arrival and reception of forced migrants. Second, it will point out the intersections between forced migrants' needs and welfare provision. Third, an overview will be sketched out of the restructuring of social policy, ongoing processes of rescaling and the deriving impacts on the European welfare systems (Brenner, 2004; Kazepov, 2010; Ranci et al., 2014). This literature will provide a theoretical background to inform similar patterns in migration, but also reception policy. The methodological section will follow. Finally, the paper will explore the specific system of reception in Italy alongside the evolving division of responsibilities and the increasing process of rescaling down to local authorities. It will do so by drawing from empirical evidence collected through 5 research works carried out by the author in Northern Italy (see tables), in the period 2014–2020.

## The role of local authorities in migration policymaking

Forced migrants have long been considered as a temporary population. As a result, related policy solutions have been temporary in nature. The normative assumption of temporariness has justified, for example, the confinement of forced migrants in refugee camps (Kibreab, 2007). Besides, according to Darling (2017), it explains why social scientists have started investigating the asylum system, from an urban perspective, only in the last two decades.

In the specific field of geography, a few strands of research have explored the experiences of forced migrants in the Global South and their precarious legal status; forced migrants' dispersal, arrival and settlement in the Global North; cities as sites of border controls; sanctuary cities; urban informality and presence (ibidem). Beyond the field of geography, scholars with other academic backgrounds, from anthropology, to sociology, to political science, to welfare (more generally on migration: Caponio & Borkert, 2010; on forced migrants: Glorius & Doormernik, 2020, 2020; Campomori & Ambrosini, 2020; Agier, 2018; Kreichauf, 2018; Sanò, 2018; Pinelli 2017; Fassin, 2005), share the view that the polis, the urban context is a crucial site for forced migrants' experiences, as will be explained, although the relevance of the urban dimension is not always explicitly theorized.

Cities shape the experiences of migrants and vice versa. Migrants and forced migrants are part of the urban population. In cities they become potential beneficiaries of urban services. Local authorities, on their side, are faced with their presence, including arrival, settlement and onward movements and have to address their needs, while also finding policy solutions to cope with them. How do they do it? What are their specific responsibilities? How do they share them with other government agencies at different government levels?

Within the wider debate on migration governance, social researchers and political scientists have described local policy interventions as driven by a "pragmatic logic" (on Europe: Caponio & Borkert, 2010; Jørgensen, 2012; on Italy: Marzorati et al., 2017). Local authorities are challenged by the presence of concrete people with concrete needs (Ambrosini & Boccagni, 2015; Vermeulen & Stotijn, 2010).

Local administrators have to take local political circumstances into consideration, including voters' demands for safety and attitudes of closures toward foreigners. Generally speaking, migration-related policy interventions have a very strong symbolic dimension and tend to be very contentious as a result (Della Porta, 2018).

Their contentious character is often more evident with regard to discourses. Approaches to policy interventions can diverge from policy discourses, particularly during electoral campaigns (Garcés-Mascareñas & Penninx, 2016). Implementation can also differ because of street-level bureaucrats' discretion (Lipsky, 2010) and individualized application of professional norms (Dubois, 2016) which can lead to differing practices for (perceived) "deserving and less deserving" beneficiaries (on the Netherlands: Van der Leun and Bouten, 2015; on the US: de Graauw, 2015; on Italy: Marzorati et al., 2017; on the EU: Ambrosini, 2018).

In relation to deserviness, Fassin (2005) strongly argued on the increasing confusion between what he calls a "humanitarian government", characterized by the deployment of moral sentiments toward "poor people" and the political. In his view, the humanitarization of the management of asylum seekers has led to a new relation between the camp and the polis: "*Because these regimes defend the polis for the happy few, they invent the camp for the undesirable. In the former, life is recognized as the political existence of the citizen, whereas in the latter, it is reduced to the bare life of the vagabond.*" (381)

Local authorities have a crucial role but do not operate in isolation (Glorius et al., 2019). In face of excluding national policies, some authorities have put in place services for irregular migrants, who are usually not granted state social protection (Ambrosini, 2018; Varsanyi, 2006); others have granted reception to (otherwise stranded) asylum seekers or refugees (Oomen et al., 2016). Differently, some authorities have designed interventions that reinforce excluding national policies (Ambrosini, 2012; Chand & Schreckhise, 2015; Gargiulo, 2017), sometimes under the pressure of national governments to reduce alleged pull factors for migration (e.g. easier access to welfare services) (O'Mahony and Sweeney, 2010).

Scholten (2013) identified four ideal-types of governance configurations: centralist (top-down), localist (bottom-up), multilevel, and decoupled. The former displays a hierarchical division of responsibilities between government levels; the localist type stresses the relevance of the "local dimension" (Caponio & Borkert, 2010) and the specific responsibility of local governments in dealing with migrants' needs; the multilevel type is characterized by a dynamic coordination across government agencies and levels; the decoupled type describes the absence of any meaningful policy coordination among levels.

While the framework provided by Scholten has the merit of encouraging exploration of layers of government from the European down to the local level, it is focused on forms of coordination, while overlooking the dimension of contention that is particularly evident in Italy (Marzorati et al., 2017; Campomori & Ambrosini, 2020; Semprebon and Caroselli, forthcoming), but not only, as the struggles of sanctuary cities show well (Darling and Bauder, 2019).

Additionally, we need to account for the fact, that notwithstanding governance configurations and contentious dynamics associated with them, national governments still bear the primary responsibility for immigration policy, while local governments are responsible for (welfare) service delivery. At the same time, a shift of responsibilities have been observed over the last decades, in line with the wider field of social policy, as it will be discussed and they have been characterized themselves by dynamics of contention. On the one hand, national governments have shifted their responsibilities on migration up (i.e. toward international and supranational institutions), out (i.e. toward nonpublic actors, but also foreign countries for the externalization of borders) and down (i.e. toward local-level authorities) (Van der Leun and Guiraudon & Lahav., 2000; Van Der Leun & Bouter, 2015); on the other, international institutions, local authorities and civil society organizations have strived to have their say on and influence national migration policies (Scholten, 2011).

## The intersections between forced migrants' welfare rights the role of the city

As anticipated, (welfare) service delivery is the primary responsibility of local governments. This is why the right to welfare services has been strongly connected to Lefevre' notion of "inhabitance", connected to the quotidian dimension of living in a locality. On the weave of a growing literature on urban citizenship and the right to the city, Purcell (2006) warned on the risk that a similar approach to welfare rights may lead to prioritizing the right of local residents over those of the wider public. Access is not universal and by definition it has always been contentious. Normally it is only the resident members of the population who can enjoy full access to provision, although EU Member States have incorporated the social rights of resident citizens, resident non-citizens and nonresidents in different ways (Sainsbury, 2012). A complex scenario of diversified welfare entitlements has derived (Spencer, 2018) and this has become even more evident in Italy in the last decade, and particularly in 2011, and later in 2015–2016, with increasing arrivals of asylum seekers on Southern Italian coasts and through the borders with Austria, France and Slovenia and the increasing need to provide reception services.

Dispersal is a policy field with considerable challenges for local governments. It is part of a distinct policy framework, composed of international conventions, European and national laws, but it is also strictly connected to welfare policy. The specific legal status of migrants affects their eligibility for welfare services. Limited access to them contributes to the stratification of social rights that in turn affects the economic and social well-being of migrants and their position in the society in the short, medium and longer term (Sainsbury, 2012).

The Geneva Convention provides a general frame for refugees' social protection: they must be provided the most favorable treatment possible, "not less favorable than that accorded to aliens" (art. 7).

Generally speaking, European Member States provide more restricted access to services for asylum seekers (and undocumented migrants) compared to refugees (e.g. Ambrosini, 2018; Spencer, 2018). The specific conditions of other migrants, such as denied asylum seekers, but also holders of subsidiary and humanitarian protection, have been less explored in social protection research. Yet, each of these statuses involves critical issues, including first of all the conditional role of the legal status itself (Corrigan, 2014; Shutes, 2016) on the changes of inclusion (Könönen, 2018).

In this regard, the service delivery role of local authorities is crucial, as we shall see later with specific reference to the Italian case. This calls again for improved attention to the division of responsibilities across agencies and the tensions that have characterized it, in the attempt to avoid any methodological nationalism or localism (Brenner, 2009). The question remains open: if the local authority does not deliver services to specific groups of migrants, who does? Who should be held accountable for this?

## The restructuring of social policy, processes of rescaling and the impacts on European welfare systems

The European Union has promoted an agenda of broadening social rights, including also the social protection of immigrants (Soysal, 1994). At national level, political debates have heated up to define the list of legitimate welfare receivers, while some political movements have insisted on presenting immigrants as a "burden to welfare states" (Albertini and Semprebon, 2018).

In this apparently contradictory scenario, a gradual process of reforms has taken place, and is still ongoing. It is a process characterized by the territorial re-organization of social policies, the multiplication of actors involved in the planning and implementation of welfare interventions (Brenner, 2004) and the growing role of the nonprofit sector in the implementation of services (Marzorati et al., 2017).

Following various waves of restructuring (Kazepov, 2010; Rauch, 2008) and growing pressures toward the devolution of responsibilities, welfare scholars have explored the (changing) relationships and dynamics between governmental actors at different scales. A particular focus has concerned processes of rescaling, both upwards, from the national to the European Union level, and downwards, toward regional and local authorities (Ferrera, 2005; Kazepov, 2010).

Processes of rescaling have derived, on the one side, from local authorities' demand for more autonomy and, on the other, from nation-states implementation of decentralization reforms, in face of the fiscal crisis and the inadequacy of national welfare programs in coping with the new emerging social risks (Ranci et al., 2014).

Most European countries have experienced these processes, but with different timing and speed. Impacts have been stronger in Southern European countries, where the division of responsibilities among different institutional levels is not clearly defined and is characterized by overlapping competences (as well as loopholes and ambiguities) (Sabatinelli and Semprebon, 2017). Southern European countries, such as Italy, are characterized by scarcely inclusive social protection systems: limited subjective rights and minimum standards only for health – not social – services. Hence, the restructuring of social policies has contributed to the increase of preexisting territorial divides, across but also within regions (Pavolini, 2015), thus threatening the principle of equality and putting territorial cohesion under risk (Kazepov & Barberis, 2013).

As far as downward rescaling – which is most relevant to this paper – is concerned, the decentralization of social policies can favor the emergence of innovative and context-specific policy solutions. However, downward rescaling can also entail a less transparent system of responsibility and reduced accountability (Bonoli, 2012; Keating, 1998), as well as a gradual delegation of responsibilities to nonpublic actors. This often responds to the aim of reducing public expenditure, although it tends to be justified by a rhetoric of subsidiarity. What is more, in situations of austerity, responsibilities have been often transferred to the local level without transferring adequate resources to implement policies. Local government agencies have been left to raise them autonomously.

## Migration, reception policy and rescaling

The governance of migration has built on the legacy of social policy, including the trends that have characterized it. Hence, the processes of rescaling described above should be taken into full account. At the same time, the governance of migration has also developed features of its own.

On the one side, researchers have analyzed the processes associated with the territorial dimension of migration policy (Glick Schiller and Caglar, 2011) and the role played by various actors (Caponio & Borkert, 2010; Caponio & Jones-Correa, 2018), from national to subnational ones. An increasingly relevant role has been recognized to the third sector (on Italy: coauthor et al. 2017; on the United States: De Graauw, 2015) in delivering services in support of migrants (e.g. language courses, basic health services, food, clothing, shelter, advocacy and legal advice) (Ambrosini, 2018; Garkish et al., 2017). Local government agencies often outsource the implementation of services to the third sector, as far as immigrants are concerned. They have repeatedly delegated full responsibility to deal with undocumented migrants and asylum seekers whose application is rejected (Ambrosini, 2018). In compliance with national legislation, they would otherwise not be catered for (coauthor et al. 2017; Van der Leun and Van Der Leun & Bouter, 2015). According to some interpretations, the third sector is de facto acting as a substitute of the public sector (on the United Kingdom: Mayblin and Poppy, 2019; on Italy: Ambrosini, 2018; Balbo, 2015), particularly at a time of austerity, in an actual rhetoric of exclusion (Castañeda, 2007).

On the other side, researchers have looked at rescaling by investigating the growing importance of local government agencies not only in the implementation of national normative frameworks but also in the development of their own policy intervention. This suggests that the traditional distinction between migration policy and immigrant (integration) policy, and the related competences of the national or local government agencies, has become more and more inadequate (Barberis & Violante, 2013). An evident example in this sense is provided by policies that have regulated (and restricted) access to residency permits (on the basis of additional criteria) (Rigo, 2011; Ambrosini, 2012; coauthor et al., 2017; Gargiulo, 2017).

Turning to asylum seeker reception, various authors have stressed how implementation has been largely – although not fully (specific details on Italy will be given later) –

| TITLE OF PROJECT | PERIOD | FOCUS | INTERVIEWS/ | LOCATION | PUBLISHED MATERIAL |
|---|---|---|---|---|---|
| PRIN National project *"Piccoli comuni e coesione sociale. Politiche e pratiche urbane per l'inclusione spaziale e sociale degli immigrati"* Available at: http://www.unescochair-iuav.it/en/research/prin-small-size-cities/ Collaboration as researcher of Milan-Bicocca University (Milan) (Funding: Italian Ministry of Education and University) | 2013–2015 | Governance of migration and reception in small municipalities | 94 (local politicians, municipal officers, associations and volunteers, reception stakeholders) | Lombardy Region: small municipalities in provincial area of Brescia, Pavia, Mantova | Marzorati et al., 2017 |
| FAMI National project *"Capacitymetro"* Available: http://www.unescochair-iuav.it/en/capacitymetro/ Collaboration as researcher of SSIIM Unesco Chair (Funding: Italian Ministry of Education and Research – AMIF Fund 2014–2020) | 2017–2018 | Governance of migration and reception in metropolitan area of Venice | +40 (municipal officers, associations in the field of migration) | Veneto Region: Venice and small municipalities in provincial area | SSIIM Unesco Chair 2018, 2018a |
| Project *"Antenne Migranti"* Available at: https://www.alexanderlanger.org/it/948/3977 Collaboration as volunteer of Langer Foundation Stiftung (Bolzano) and volunteer unpaid researcher (Funding: Open Society Foundation) | 2016–2019 | Governance of transit migration along the Brenner route | 9 (municipal officers, reception stakeholders) | Veneto Region: Verona; Trentino-Alto Adige Region: Trento, Bolzano, Brenner | Pelacani & Semprebon, 2020 Semprebon & Caroselli, forthcoming |
| Project *"Casa Iride"* (Final report of project) Available at: https://www.spazioapertocoop.it/projects/comunita-di-accoglienza-casa-iride/ Collaboration as researcher of Cà Foscari University (Venice) (Funding: Fondazione S.Zeno, Verona) | 2019 | Local reception project and inclusion of women asylum seekers in Verona | 26 (reception beneficiaries, internship and training employers, reception operators) | Veneto Region: Verona | Della Puppa et al., 2020 |
| EU *"INSigHT Action"* (Building Capacity to Deal with Human Trafficking and Transit Routes in Nigeria, Italy, Sweden) Available at: https://www.insightproject.net/ Collaboration as research coordinator and researchers of SSIIM Unesco Chair (Funding: EU – DG Home; Managing Agency: ICMPD) | 2019–2021 | Governance of protection and reception of women victims of trafficking in the Veneto Region | 66 (antitrafficking and reception stakeholders; beneficiaries) | Veneto Region: Venice, Verona, Padova | Semprebon et al. forthcoming |

| TITLE OF PROJECT | INTERVIEWS USED IN THIS PAPER |
|---|---|
| PRIN National project *"Piccoli comuni e coesione sociale. Politiche e pratiche urbane per l'inclusione spaziale e sociale degli immigrati"* | managing director SPRAR/CAS structures, Brescia, 13/05/2019<br>managing director of SPRAR/CAS structures, Breno, Brescia, 27/03/2014<br>mayor, Malegno, Brescia, 07/03/2015<br>coordinator of Reception Network in Lombardy Region, 28/03/2014 |
| FAMI National project *"Capacitymetro"* | mayor, Cona, Venice, 06/06/2018<br>chief officer Social Services Department Municipality of Mirano, Venice, 24/03/2018 CAS operator, Venice, 13/07/2018<br>chief officer Immigration Office, Municipality of Venice, 20/11/2017<br>mayor Artogne, Brescia, 26/03/2014 |
| Project *"Antenne Migranti"* | CAS operator, Verona, 20/08/2019<br>socio-legal operator CIR, 15/03/2018<br>CAS operator, Sanguinetto, Verona, 15/11/2017 |
| Project *"Casa Iride"* | – |
| EU *"INSigHT Action"* (Building Capacity to Deal with Human Trafficking and Transit Routes in Nigeria, Italy, Sweden) | social assistant, Municipality of Verona, 27/05/2020<br>officer Social Services Department, Venice, 26/05/2020<br>chief officer Reception Office, Municipality of Verona, 25/06/2020 |

outsourced to private and third sector actors who have been entrusted with responsibility to manage reception structures and deal with the needs of beneficiaries, with limited accountability on the side of local entities. At the same time access to reception structures has been organized on the basis of a system of dispersal, based on quotas to "share the burden" (among others for Italy: Semprebon and Pelacani (2020); Robinson et al., 2003). In face of these features, critics have been raised as will be explained, suggesting that forced migrants have not been granted adequate social protection.

Darling (2017) criticized dispersal and reception programs for the devolution of responsibility for forced migrants' protection. Reception includes a "bundle of specific welfare services" (accommodation, food, educational/training activities), whose responsibility is transferred from central government to local actors, through channels that are parallel to those used to provide social protection to residents and resident migrants. Taking this argument further, this paper warns that a similar approach to the governance of reception has contributed to the consolidation of a parallel, exceptional system of welfare that hardly allows for the inclusion of forced migrants, particularly in the long-term.

The devolution of responsibility for reception follows a pattern that seems coherent with the rescaling of social policies. Dispersal has been often narrowly understood as the mere (physical) allocation of asylum seekers within the nation state (Boswell, 2001; Robinson et al., 2003) and it has been largely undertaken in a logic of control, rather than participation (Griffiths et al., 2005; Boswell, 2001), thus contributing to the marginalization (Schuster, 2004) of a population that is still considered as temporary. Moreover, reception has been characterized by a proliferation of private and third sector actors (Darling, 2017), resulting in the lack of service standardization and in the fragmentation of provision (Marchetti, 2016; Darling, 2016; Semprebon and Pelacani, 2020), with a likely impact for asylum seekers in terms of equality of access,

## Methodological framework

This paper will build on empirical evidence collected through five research works undertaken by the author in Northern Italy, in the period 2014–2020. The time frame coincides with the second main wave of arrival of asylum seekers in Italy (2015–2016) and the reception that was organized afterward, but also before it. In terms of location, the paper is biased as it deals with North-eastern Italy, where the ordinary system of reception has been less present (the majority of structures of in the South,[1] hence in cannot be generalized for the whole country.

All the five works focused on the reception of asylum seekers, as the tables below detail:

Evidence of these research work will be provided mainly by referring to published material, but also by providing the reference of some of the most significant interviews undertaken on the specific focus of this paper.

As anticipated, in order to explore the division of responsibilities within the system of reception, while also highlighting how such system has contributed to the development and strengthening of an exceptional parallel system of welfare for forced migrants, the paper will look particularly at scaling and rescaling processes to move beyond the framework by Scholten. It will explore the tensions and forms of contention and not only the coordination that has characterized the activation of actors at different territorial levels. This will be done by looking at the specific functions and phases of reception policy: funding, implementation, coordination and monitoring.[2]

Before proceeding with the analysis, the next section will provide a short overview of the main features of the reception system.

## The Italian system of reception: a critical analysis of the division of responsibilities

### *The ordinary and extraordinary system of reception: SPRAR/SIPROIMI, CAS*

As explained in more details in Semprebon and Pelacani (2020), Italy has been characterized by a double-track dispersal and reception system, an ordinary and an extraordinary one. The origins of the ordinary system can be traced back to the National Asylum Plan that was institutionalized by Law 89/2002 under the name of SPRAR (Protection system for asylum seekers and refugees).

In 2011, a second system was set up, to cope with an increasing flow of migrants, associated mostly with the Arab Springs. Extraordinary Reception Centers (CAS) were established, under the responsibility of the Civil Protection Department, as a temporary solution addressing the limited accommodation capacity of the ordinary system. Over time, this solution was consolidated and eventually institutionalized in 2014. It is now Prefectures[3] that are responsible for CAS and particularly for the public tender procedure required to outsource the management of reception structures.

As defined by legislative decree 142/2015, asylum seekers are accommodated in SPRAR centers and, when available places are insufficient, in CAS, according to a system of progressive quotas, based on a proportional percentage of each region's inhabitants and of the resources provided by the national fund for social policy.[4]

The SPRAR consists in a network of reception projects and reception centers, mostly managed by social cooperatives. Since its creation, the network has been growing considerably: from 1,365 places in 2003, to 35,650 in January 2019[5]. According to 2019 data, the beneficiaries hosted in SPRAR reception centers included mostly single adult men (83.5%), but also unaccompanied minors (9.4%) and family units (18%). Importantly, 20.5% were indicated as individuals with vulnerabilities, stressing the fragility of their situation.[6] SPRAR projects include "integrated" reception services aimed at favoring beneficiaries' autonomy (room and board, health assistance, but also language-cultural mediation, multicultural activities, support for job seeking, legal assistance). CAS are also generally managed by social cooperatives, but normally with lower minimum standards of services with respect to SPRAR structures. Accordingly, a few authors have highlighted the existence of inhomogeneous standards of reception facilities across the country and different typologies of structures (Bolzoni et al., 2015; Marchetti, 2016; SSIIM Unesco Chair, 2018a).

SPRAR and CAS can host asylum seekers and asylum seekers that have either been granted or denied international protection. However, a trend developed in 2017, according to which priority access to SPRAR centers was given to holders of international protection and unaccompanied minors.[7]

***The division of responsibilities in the Italian system of reception: the critical role of local authorities***

### *Funding*

The SPRAR network has been mostly funded by the Ministry of Interior, hence the central government, through the National Fund for Asylum Policies and Services (FNPSA), established with the same law that institutionalized the SPRAR, and a few other European funding sources, including the European Refugee Fund.[8] This testifies a dedicated funding provision was made for reception since the institution of the network. Such provision has been provided in the past two decades. However, for the period 2008–2013, 42.8% of the funding available through the European Refugee Fund were used for "emergency" measures, associated with increasing flows of forced migrants. Additional "emergency" streams of resources were provided by the Ministry of Interior, through a series of 20 decrees, issued by the President of the Council of Ministers, in the period 2011–2013. It was in this period that reception started being framed as an "emergency" (Marchetti, 2012), thus diverting resources away from the consolidation of the SPRAR that was aimed to become the main ordinary system of reception in Italy.

The SPRAR network grows larger and could count on national funding support for almost two decades, with recent changes that have made it more precarious and susceptible to changing government coalitions – with diverging approaches to migration and reception -, thus testifying the strong contentious nature of reception as well as migration issues more generally. In 2018, the so-called Salvini Law 132/2018 abolished it and created a new network the SIPROIMI (System of Protection for the Holders of International Protection and Unaccompanied Minors), which differs from the SPRAR as it provides access to refugees, holders of subsidiary protection and unaccompanied minors only, while asylum seekers are destined to CAS. The recent changes to the Salvini

Law, approved on October 5th, (among other things) modified it to ensure asylum seekers' access to the ordinary system of reception.

Funding was provided also in the form of a "reception bonus", introduced with Decree 25 January 2017, that is to say a monetary incentive, targeted to municipalities, in face of increasing complaints of local governments having to keep "calm down" protests against asylum seekers arrival, particularly in locations with very large reception structures. Bonus were given to municipalities upon formal confirmation they would not oppose the arrival – in higher numbers with respect to other municipalities – of asylum seekers. These incentives were meant to compensate local communities for their acceptance of national government's devolution of responsibility with regard to the implementation of reception, although de facto responsibility has been resting in the hands of Prefects. Paradoxically, such bonus was generally used for maintenance of local roads and similar interventions, rather than as extra financial resources to improve the conditions of asylum seeker reception.[9]

Looking at specific structures, an average funding of 30–35 euro was granted for each beneficiary hosted, to cover all costs (accommodation and board, psychological and legal support, Italian language courses and a small pocket-money for beneficiaries' personal expenses). Following the Salvini Law, a severe cut in expenditures was implemented: the funding was brought down to 20 euro per beneficiary,[10] with reception operators claiming this will not allow an adequate standard of reception, but only the provision of basic services, far from the idea of granting effective social protection to beneficiaries and favor their inclusion.[11] This is particularly evident for CAS that were no longer required to provide Italian courses, nor psychological support, which were both left to the discretion of the operators in each CAS. It is a clear example of how the rescaling of reception responsibility out to social cooperatives has followed more and more a logic of devolution without resources, leaving local level actors alone in deciding how to deal with it or whether to close down services.[12] There are in fact cases of social cooperatives that have decided to close down reception.

### Implementation

The main requirement for the activation of a SPRAR/SIPROIMI project is to have a local entity applying for funding and taking responsibility as the main signatory. However, applications are voluntary and while the number of local entities applying has been growing, the vast majority of them has not done so. As of March 2019, approximately 1,800, out of a total of more than 8,000 in the whole country, had applied.[13] This suggests scarce engagement on the side of local authorities. This partly explains why the extraordinary system has been prevailing, in absolute number of places. More involvement is evident in larger cities, where it can be arguably expected that more human and economic resources are available. At the same time, a more relevant variable seem to be the strong long-term engagement of municipalities with immigration and reception, not only in larger cities, but also in smaller ones.[14]

Levels and modalities of collaboration between public administrations and third sector actors have varies across the country, across regions but also within the same region and/or local territories: cases have been recorded of public administrators merely signing the project, while outsourcing the full implementation to a social cooperative, with a very minimum quota of co-funding, as it happened in many cases[15]; but also of public administrations refusing to sign for a SPRAR project, in face of local communities' resistance as well as the fact that migrants' arrival is often communicated a few days before arrival without no possibility of planning and organizing it[16]; to public administrations being fully involved with co-funding in the form of human and economic resources, because of a legacy of engagement with migration and reception issues, or even by the existence of a dedicated office and undoubtedly also more available resources, as a larger municipality.[17]

This is particularly significant if we recall the fact that the Italian Constitution (art. 18) spells out the duty, for each local authority, to deal (not based on a voluntary choice) with the needs of the people who live on their territory, although not residents of it. This is part of a broader range of (administrative) responsibilities,[18] it is linked to the wider social and welfare policy framework, for which local authorities are held responsible (i.e. public health protection, housing, community cohesion, etc.). The framework is relevant for migrants too depending on their status, notwithstanding the fact that cities have often resorted to address, often with the help of the third sector, the needs of all people on their territory, including undocumented migrants (Ambrosini, 2018) who only have minimal welfare rights.

As regards CAS structures, the actual management of the projects has been generally outsourced to social cooperatives, with proliferation of actors has been observed, particularly social cooperatives, often with headquarters in other Italian regions, unknown to local actors, with limited professional experience and no knowledge of the local territory.[19] The shortage of professional experience had been already reported by the SPRAR, particularly as efforts were made to enlarge the network,[20] which required a rapid introduction of human resources.[21]

Generally speaking, an emergency approach has been rather pervasive. Not only has the public sector scarcely engaged in reception. It has engaged even less in post-reception for holders of international protection. In fact, hardly any policy, with the partial exception of AMIF (Asylum and Migration Fund) projects, has been planned nor implemented for holders of international protection, to support them in their inclusion, with the result they run the risk of being stranded upon completion of their reception period, as local authorities are not always willing to follow up on any remaining needs, although associated with vulnerabilities, having to deal with limited municipal budgets.[22]

As a further result of the Salvini Law, humanitarian protection[23] was abolished and holder of humanitarian protection could not renew their permit to stay. This threatened to impact even harder on their post-reception trajectory. Some municipalities recurred to temporary solutions, such as shelters for homeless people, that are already under strain,[24] and that often require residency status for people to access them. The COVID-19 pandemic further affected migrants,[25] but exploration of this aspect goes beyond the scope of this paper.

A point on residency status should also be made about asylum seekers, as the Salvini Law introduced new extra requirements to obtain it. As suggested by ASGI,[26] such requirements are unconstitutional. Yet they have been legitimizing discriminating practices – that were already under way before the passing of the Law – by local authorities: many local authorities refused to (or delayed) access to residency although it is a right for asylum seekers as spelt by decree 142/2015[27]. While this does not prevent asylum seekers from accessing the reception system, it does create difficulties in obtaining the national health card and in accessing a number of other welfare services, including access to social housing, following the completion of the reception project. The result is a differential "rescaled" inclusion of migrants not only based on citizenship, but also based on residency status (see Gargiulo, 2017), pointing to the fact that the distinction between migration policy and immigrant (integration) policy, and the related responsibilities, are more and more entangled and yet fragmented (see Barberis & Violante, 2013). This sheds light on how the lack of engagement of local authorities in reception can in fact be an engagement in producing barriers to justify the absence of adequate welfare policy provisions.

The welfare needs of forced migrants have often been dealt by local authorities as exceptional needs that are beyond their responsibility – either because of implementation was outsourced to social cooperatives or because of more or less evident residency obstacles in the post-reception phase. Arguably this is precisely what is contributing not only to the consolidation of an emergency system but also to a parallel, exceptional system of welfare, particularly with reference to the post-reception phase.

### *Coordination and monitoring*

The SPRAR/SIPROIMI involves a mechanism of coordination, entrusted by the Ministry of Interior onto the SPRAR/SIPROIMI Central Unit that provides technical support to local authorities and social cooperatives, promotes the network, proposes trainings to operators and organizes monitoring activities (see SPRAR website). The CAS system has followed a different route, characterized by an increasing privatization of reception projects, with a general lack of coordination. This can be understood as an outcome of what was established as a secondary, emergency system, to provide a temporary solution to the shortage of places in the SPRAR. There are some exceptions.

Ad hoc coordinating tables have been organized whereby the Prefect was willing to do so. For example, thanks to the engagement of an Officer or the Prefecture of Venice, a local network of CAS structures was created, as early as 2014[28]. Initially regular weekly meetings were organized with the main intention to cope with the "emergency", that is to say the increasing arrival of migrants. At a second stage, engagement improved toward the definition of a common vademecum to share good practices in the management of CAS structures and specific structures were activated to deal with asylum seekers with special needs. In the nearby Milanese conurbation, another network was set up, following the engagement of operators of CAS and SPRAR structures, with the main goal of facilitating the exchange of information and the provision of reciprocal support in the management of reception structures and in dealing with bureaucratic issues.[29] The fact should not be underestimated that what is still missing for CAS structures is a coordinating unit at a higher, central level, with functions similar to the SPRAR/SIPROIMI Central Unit.

One criticism that can be addressed to the SPRAR/SIPROIMI is the opacity of the system of dispersal (coauthor et al. 2020; see also Biehl, 2007; Sorgoni, 2011; Whyte, 2011): no information are given in any SPRAR reports of the criteria adopted by the Central Unit to proceed with dispersal, apart from the fact that vulnerable asylum seekers are transferred to centers that have adequate resources and tools to cope with their specific vulnerability. On their side, asylum seekers have no space for their self-determination. They cannot choose where to be transferred, although they are likely to have some preferences, based on existing networks or considerations linked to employment opportunities (see Boswell, 2001; Griffiths et al., 2005). Interestingly, most SPRAR projects are based in central and southern Italy, while most job opportunities are concentrated in the north (see SPRAR website for more information).

Opacity in the system of dispersal is a characteristic of CAS too. Prefectures have rarely involved municipalities whenever activating a reception project, with the result that mayors and local communities have sometimes felt reception as imposed onto them. This was particularly evident for smaller and more isolated municipalities with less economic resources and less available structures. Emblematic cases are those of Cona and Artogne – that also hit the headlines of national newspapers, where asylum seekers were transferred to very large reception structures entirely managed by a social cooperative, as agreed with the Prefecture with no prior confrontation with the mayors.[30] It must be specified that the relatively recent responsibility for CAS entrusted onto Prefectures seems to be hardly adequate. It is reasonable to wonder how the inclusion of forced migrants could be favored by an institution that has been traditionally responsible – and has developed specific competences – to ensure order and safety within the provincial territory and whose collaboration with local authorities and particularly social assistants used to be limited to this specific task.

**Reception policy in Italy: a parallel exceptional system of welfare and the potential impact on forced migrants**

Following Scholten's classification, the Italian system of reception can be associated with a "localist" model, although hints of its "decoupling" nature are also evident if we focus on the scarcity of policy coordination among actors at local and higher territorial scales. However, this paper argues that for a more detailed analysis of the reception system in Italy more attention must be focused on the allocation and rescaling of specific responsibilities, in line with processes in the wider field of social policy. Besides the contentious character of such dynamics should be highlighted as it allows further understanding of the functioning of the whole system of reception and the potential impact on post-reception and the social protection of forced migrants.

The Italian system of reception has been evidently operated on the basis of an emergency approach, as the trends in funding show, as well as the prevalence of available places in the CAScompared to SPRAR structures.

The implementation of reception has been largely left in the hands of Prefects who have rescaled them out to private and third sector actors but with scarce transparency and limited accountability. A "bundle" of services has been provided to forced migrants through an opaque fragmented and discretional system, depending on individual reception structures if not individual operators. This has resulted in a gradual retrenchment of welfare provision, even more so following the Salvini Law, the reduction of funding and the lowering of minimum standards, besides the restriction of asylum seekers' access to to the CAS only. All this with hardly any consultation of local authorities.

Prefects have been entrusted with the responsibility to identify and select CAS structures, but have often called for local authorities to find them, although without any direct involvement in the procedure. At the same time, through the introduction of a "compensating bonus", local authorities have been encouraged to accept the arrival of forced migrants on their territory, while "keeping under control" increasing protests by local communities. This shows that local authorities have had a role in reception, although not clearly defined.

The scenario changes if we focus on the SPRAR system, that requires the formal partnership of a local entity. The local authorities that have signed have portrayed different level and forms of engagement. Those that have engaged the most are those characterized by a stronger legacy of activation in migration issues. It remains strikingly evident that the majority of local authorities have not joined a SPRAR project at all and the voluntary option of joining has probably contributed to discouraging it. In line with the argument by Fassin (2005), most local authorities have concentrated on "defending the polis for the happy few, while accepting camps for the less desirable citizens" and not taking responsibility for them.

But what happens when forced migrants complete their reception project? The SPRAR was institutionalized with a vision of inclusion, aiming to facilitate asylum seekers' autonomy and self-determination. As stated in many SPRAR Annual Reports, however, it is only some beneficiaries that manage to find decent housing, a regular job and eventually settle down. Some encounter difficulties associated with poverty and/or endure forms of racism and discrimination (Malkki, 1995; Zetter, 2007). Some find themselves stranded (Médecins Sans Frontières, 2016) in a state of "permanent temporariness" (Vosko et al., 2014), as testified by squatting experiences (e.g. Médecins Sans Frontières, 2016 – on similar experiences in other EU countries see Griffiths et al., 2005; Squire, 2009). The situation in CAS can hardly be expected to be better, as explained. Hence, when beneficiaries terminate their reception project, it is mainly to local authorities they could turn to for access to welfare services, but local authorities are reticient in this sense, partly because of their unclear role in reception (see also Marchetti, 2014), partly because of the contentious definition of entitled welfare receivers – that will draw resources from local governments – not only on the basis of citizenship but also residency status. Hence in the role of local authorities the traditional distinction between migration policy and immigrant (integration) policy, has becomes more and more blurred. De facto it is local authorities that ultimately decide to what extent they respond to welfare needs. In this sense, reception localities have become "internal borders" (Fontanari & Borri, 2017), where "local" bordering processes take place (Lebuhn, 2013) to define the legitimate community of welfare receivers. Arguably, local authorities could be more prone to take up responsibility in post-reception phase whereby they are increasingly involved throughout the process of reception, as argued by coauthor et al. (2019). But this would require coordination at local level, that is evident but only in some casese, particularly whereby policy entrepreneurs take the lead (coauthor et. al, 2017).

Generally speaking, this paper has tried to show how the functioning of the system of reception in Italy has led toward the consolidation of a parallel exception system of welfare for forced migrants that hardly allows for the inclusion, particularly in the long-term. The COVID-19 pandemic has been this even more evident.

As Könönen (2018) suggests, the regulation of entitlements should be examined in its entirety because of the strong interrelations between legal status, access to employment and social rights and more importantly because (the regulation of) rights concerns the indivisible life of citizens as well as non-citizens (Könönen, 2018). Arguably, the "bundle" of services for forced migrants has shun attention away from concerns for the overall welfare of forced migrants and made their position more fragile by preventing asylum seekers to reclaim social protection as residents (including access to housing, health, etc.) and continue doing so, where necessary, upon completion of their reception project, when, on the contrary, they are often hardly aware of how it functions as they have been kept away from mainstream welfare provision.

It is also worth recalling that forced migrants, without a regular permit to stay, may in fact call for the right to go not for the "right to come and stay" (Fernandez & Olson, 2011, p. 415), thus stimulating further reflections on the universal theme of mobility (e.g. Mezzadra, 2011). More research in this promises to put under discussion welfare systems based on the principle of universal residence-based welfare and its excluding potential (Könönen, 2018). The implications for policy would include the re-definition of the "legal fictions of residence" (ibidem) for non-citizens and citizens alike and the construction of a welfare model capable of overcoming the trend to plan and implement "sedentary" policy solutions to address policy fields, such as that of migration, that by definition are international, but also internal, in nature, and characterized by flows, mobility, fluidity.

## Notes

1. See note 4 for data
2. This reflection was inspired by collaboration in the EU COST Action IS1102 Cohesion – Social services, welfare state and places » (Martinelli et al., 2017).
3. The Prefecture is an administrative peripherical body of the Italian Government, under the hierarchical responsibility of the Ministry of Interior and it operates at provicial level.
4. See the "Intesa tra il Governo, le Regioni e gli Enti locali sul piano nazionale per fronteggiare il flusso straordinario di cittadini extracomunitari, adulti, famiglie e minori stranieri non accompagnati (National Agreement between the Italian Govnerment, Regional Authorities and Local Authorities to deal with the exceptional flows of non-European citizens, adults, families and unaccompanied minors) Retrieved from http://www.prefettura.it/FILES/AllegatiPag/1247/Accordo%20Conferenza%20Unificata.pdf Last access: 3 July 2018.
5. SPRAR (2019) Atlante SPRAR 2018. Retrived from https://www.siproimi.it/pubblicazioni/atlante-sprar-siproimi-2018 Last access: 20 January 2021.
6. See note 5.
7. SPRAR (2017) Atlante SPRAR 2016. Retrived from https://www.sprar.it/pubblicazioni/atlante-sprar-2016 Last access: 12 October 2018.
8. As reported by Lunaria (2013), SPRAR and CAS have also been funded through other channels (i.e. the European Refugee Fund, the National Operative Program, a program specifically for Southern Italy). Retrieved from http://www.cronachediordinariorazzismo.org/wp-content/uploads/rights_are_not_expense.pdf
9. SSIIM Unesco Chair, 2018a; Interview, mayor, Cona, Venice, 6 April 2018.
10. Ziniti A. "Viminale, tagli dell'accoglienza per i migranti da 35 a 20 euro a giorno." Retrived from https://www.repubblica.it/cronaca/2018/11/07/news/viminale_tagli_dell_accoglienza_per_i_migranti_da_35_a_20_euro_a_giorno-211025426/?refresh_ce Last access: 1 March 2019.
11. Author and coauthor, forthcoming; Interview, managing director SPRAR/CAS structures, Brescia, 13 May 2019; Interview, CAS operator, Verona, 20 August 2019.
12. Ibidem note 9.
13. See SPRAR website for updated data. Retrieved from www.sprar.it Last access: 3 March 2019.
14. SSIIM Unesco Chair, 2018a.

15. Marzorati et al. (2017); Interview, mayor, Cona, Venice, 6 April 2018; Interview, managing director SPRAR/CAS structures, Breno, Brescia, 27 March 2014.
16. SSIIM Unesco Chair, 2018a; Seminar "L'accoglienza dei richiedenti asilo a Verona", Centro Studi Immigrazione, Verona, 16 December 2018; Interview, mayor, Malegno, Brescia, 7 March 2015.
17. SSIIM Unesco Chair, 2018; Interview, chief officer Immigration Office, Municipality of Venice, 20 November 2017; Interview, chief officer Social Services Department, Municipality of Mirano, Venice, 24 March 2018.
18. This interpretation is supported by various constitutional lawyers: see Vita.it, Interview, "Asgi: Tutti i Comuni devono aprire progetti Sprar, è scritto nella Costituzione", Retrived from http://www.vita.it/it/article/2016/05/05/asgi-tutti-i-comuni-devono-aprire-progetti-sprar-e-scritto-nella-costi/139297/Last access: 30 October 2018.
19. SIIM Unesco Chair 2018a; Sempreb0n (2017); Interview, managing director SPRAR/CAS structures, Brescia, 13 May 2019; Interview, operator, CAS Verona, 20 August 2019.
20. SPRAR (2014) Atlante SPRAR 2012–2013. Retrieved from https://www.cittalia.it/images/file/atlante_sprar_completo_2012_2013.pdf Last access: 12 October 2018.
21. Semprebon and Pelacani (2020); Interview, coordinator of Reception Network in Lombardy Region, 28 March 2014.
22. Semprebon et al. (forthcoming); Interview, chief officer Reception Office, Municipality of Verona, 25 June 2020; Interview CAS operator, Verona, 20 August 2019, Interview, managing director SPRAR/CAS structures, Brescia, 13 May 2019; CAS operator, Venice, 13 July 2018.
23. This is a specific residual form of protection granted by Italian Law 40/1998 (similarly granted by other EU States), besides the international forms of protection associated with the status of refugee and with subsidiary protections.
24. Della Puppa et al. (2020); Semprebon and Pelacani (2020); Interview, socio-legal operator CIR (Consiglio Italiano Rifugiati), partner of SPRAR project in Verona, 15 March 2018.
25. Interview, social assistant, Municipality of Verona, 27 May 2020.
26. Consoli D. and Zorzella N. "L'iscrizione anagrafica e l'accesso ai servizi territoriali dei richiedenti asilo ai tempi del salvinismo", 8 January 2019. Retrieved from https://www.asgi.it/asilo-e-protezione-internazionale/liscrizione-anagrafica-e-laccesso-ai-servizi-territoriali-dei-richiedenti-asilo-ai-tempi-del-salvinismo/Last access: access: 3 March 2019.
27. Della Puppa et al. (2020); SSIIM Unesco Chair, 2018a; Semprebon et al., forthcoming; Interview, CAS operator, Sanguinetto, Verona, 15 November 2017; Interview, social assistant, Municipality of Verona, 27 May 2020; Interview, officer Social Services Department, Venice, 26 May 2020.
28. SSIIM Unesco Chair, 2018; Interview, coordinator of Reception Network in Lombardy Region, 28 March 2014.
29. Interview, coordinator of Reception Network in Lombardy Region 28 March 2014.
30. SSIIM Unesco Chair, 2018a; Interview, mayor, Cona, Venice, 6 April 2018; Seminar "L'accoglienza dei richiedenti asilo a Verona", Centro Studi Immigrazione, Verona, 16 December 2018; Interview, mayor Artogne, Brescia, 26 March 2014.

## Acknowledgments

An earlier version of this paper was presented at the 2017 Annual IMISCOE Conference, in the panel "Refugee Migration and Urban Studies – Theoretical Challenges and New Approaches". My thank you is due to the panel coordinators Birgit Glorius and René Kreichauf for their multiple feedbacks, to the reviewers of Urban Geography and Rosa Mas Girald for her feedback and encouragement. The research on which this paper is based has been undertaken as part of the research projects indicated in the tables and thanks to the funding acknowledged. The content of this paper is the sole responsibility of the author and can be in no way considered as expressing the opinion of any of the funders or the project leaders/partners.

## Disclosure statement

No potential conflict of interest was reported by the author.

## References

Agier, M. (2018). *La giungla di Calais. I migranti, la frontiera, il campo.* Ombre Corte.

Albertini, M., & Semprebon, M. A burden to the welfare state? Expectations of non-EU migrants on welfare support. (2018). *Journal of European Social Policy, 28*(5), 501–516. First Published online Feb. 13th. https://doi.org/10.1177/0958928717754293

Ambrosini, M. (2012). We are against a multi-ethnic society: Policies of exclusion at the urban level in Italy. *Ethnic and Racial Studies, 3*(1), 136–155.https://doi.org/10.1080/01419870.2011.644312

Ambrosini, M. (2018). *Irregular Immigration in Southern Europe: Actors, Dynamics and Governance.* Palgrave Macmillan.

Ambrosini, M., & Boccagni, P. (2015). Urban multiculturalism beyond the 'backlash': New discourses and different practices in immigrant policies across European cities. *Journal of Intercultural Studies, 36*(1), 35–53. https://doi.org/10.1080/07256868.2014.990362

Balbo, M. (2015). *Migrazioni e piccoli comuni.* Franco Angeli.

Barberis, E., & Pavolini, E. (2015). Settling Outside Gateways. The State of the Art, and the Issues At Stake. *Sociologica, 9*(2). https://doi.org/10.2383/81426

Barberis, E., & Violante, A. (2013). Rescaling, governance e frammentazione territoriale delle politiche dell'immigrazione: il caso della segregazione scolastica. In Y. Kazepov & E. Barberis (Eds.), *Il welfare frammentato. Le articolazioni regionali delle politiche sociali italiane.* Carrocci: Bologna

Biehl, J. (2007). A Life: Between Psychiatric Drugs and Social Abandonment. In J. Biehl, B. Good, & A. Kleinman (Eds.), *Subjectivity. Ethnographic Investigation*(pp. 397–422). Berkeley-Los Angeles: University of California Press.

Bolzoni, M., Gargiulo, E., & Manocchi, M. (2015). The social consequences of the denied access to housing for refugees in urban settings: The case of Turin, Italy. *International Journal of Housing Policy, 15*(4), 400–417. https://doi.org/10.1080/14616718.2015.1053337

Bonoli, G. (2012). Blame Avoidance and Credit Claiming Revisited. In G.Bonoli & D.Natali (Eds.), *The Politics of the New Welfare State* (pp. 93–110). Oxford University Press.

Boswell, C. (2001). *Spreading the costs of asylum seekers: A critical assessment of dispersal policies in Germany and the UK.* Anglo-German Foundation for the Study of Industrial Society.

Brenner, N. (2004). Urban governance and the production of new state spaces in western Europe, 1960–2000. *Review of International Political Economy, 11*(3), 447–488. https://doi.org/10.1080/0969229042000282864

Brenner, N. (2009). Is There a Politics of 'Urban' Development? Reflections on the US Case. In R. Dillworth (Ed.), *The City in American Political Development* (pp. 121–140). Routledge.

Campesi, G. (2011). The Arab Spring and the Crisis of the European Border Regime: Manufacturing Emergency in the Lampedusa Crisis. *Eui Working Papers, Rscas 2011(59),* http://cadmus.eui.eu/handle/1814/19375. Accessed: 29th June 2018.

Campomori, F., & Ambrosini, M. (2020). *Multilevel governance in trouble: The implementation of asylum seekers' reception in Italy as a battleground.* Comparative Migration Studies, online, 1–19.

Caponio, T., & Borkert, M. (2010). *The local dimension of migration policymaking.* Amsterdam University Press.

Caponio, T., & Jones-Correa, M. (2018). Theorising migration policy in multilevel states: The multilevel governance perspective. *Journal of Ethnic and Migration Studies, 44*(12), 1995–2010. https://doi.org/10.1080/1369183X.2017.1341705

Castañeda, H. (2007). Paradoxes of providing aid: NGOs, medicine, and undocumented migration in Berlin. In *PhD dissertation*, University of Arizona.

Chand, D. E., & Schreckhise, W. D. (2015). Secure communities and community values: Local context and discretionary immigration law enforcement. *Journal of Ethnic and Migration Studies, 41*(10), 1621–1643. https://doi.org/10.1080/1369183X.2014.986441

Corrigan, O. (2014). Migrant Deprivation, Conditionality of Legal Status and the Welfare State. *Journal of European Social Policy, 24*(3), 223–239. https://doi.org/10.1177/0958928714525819

Dal Zotto, E. (2014). Frame dell'emergenza e migrazioni forzate. Il caso dell'Emergenza Nord Africa. In P Musarò & P Parmiggiani (Eds.), *Media e migrazioni. Etica, estetica e politica del discorso umanitario* (pp. 125–137). Franco Angeli.

Darling, J. (2016). Asylum in Austere Times: Instability, Privatization and Experimentation within the UK Asylum Dispersal System. *Journal of Refugee Studies, 29*(4), 483–505. https://doi.org/10.1093/jrs/few038

Darling, J. (2017). Forced migration and the city Irregularity, informality, and the politics of presence". *In Progress in Human Geography, Vol 41*(2), 2, 2017. https://doi.org/10.1177/0309132516629004

De Graauw, E. (2015). Polyglot bureaucracies: Nonprofit advocacy to create inclusive city governments. *Journal of Immigrant & Refugee Studies, 13*(2), 156–178. https://doi.org/10.1080/15562948.2015.1030809

Della Porta, D. (2018). *Solidarity Mobilizations in the 'Refugee Crisis': Contentious Moves.* Palgrave Macmillan.

Della Puppa, F., Gargiulo, E., & Semprebon, M. (2020). Per una critica delle politiche migratorie italiane: Fare ricerca, prendere posizione. In N. Martorano & M. Prearo (Eds.), *Migranti LGBT. Pratiche, politiche e contesti di accoglienza*(pp. 183–199). Altera Politiche e Teorie della Sessualità.

Dubois, V. (2016). *The Bureaucrat and the Poor. Encounters in French Welfare Offices.* Routledge.

Fassin, D. (2005). Compassion and Repression: The Moral Economy of Immigration Policies in France. *Cultural Anthropology, 20*(3), 362–387. https://doi.org/10.1525/can.2005.20.3.362

Fernandez, L., & Olson, J. (2011). To live, love and work anywhere you please: Critical exchange on Arizona and the struggle for locomotion. *Contemporary Political Theory, 10*(3), 415–417. https://doi.org/10.1057/cpt.2011.15

Ferrera, M. (2005). *The Boundaries of Welfare. European Integration and the New Spatial Politics of Social Protection.* Oxford University Press.

Fontanari, E., & Borri, B. (2017). Introduction. Civil society on the edge: Actions in support and against refugees in Italy and Germany. *Mondi Migranti, 3*(2017), 23–51. https://doi.org/10.3280/MM2017-003002

Garcés-Mascareñas, B., & Penninx, R. (2016). *Integration Processes and Polices in Europe. Contexts, Levels and Actors.* Amsterdam.

Gargiulo, E. (2017). The limits of local citizenship: Administrative borders within the Italian municipalities. *Citizenship Studies, 21*(3), 327–343. https://doi.org/10.1080/13621025.2016.1277982

Garkisch, M., Heidingsfelder, J., & Beckman, M. (2017). Third Sector Organizations and Migration: A Systematic Literature Review on the Contribution of Third Sector Organizations in View of Flight, Migration and Refugee Crises. *VOLUNTAS: International Journal of Voluntary and Nonprofit Organizations, 28*(5), 1839–1880. https://doi.org/10.1007/s11266-017-9895-4

Glick Schiller, N., & Çağlar, A. (2011). *Locating Migration: Rescaling cities and migrants* (pp. 23–41). Cornell University Press.

Glorius, B., & Doormernik, V. (2020). *Geographies of Asylum in Europe and the Role of European Localities.* IMISCOE Research Series.

Glorius, B., Oesch, L., Nienaber, B., & Doomernik, J. (2019). Refugee Reception within a common European asylum system: Looking at convergences and divergences through a local-to-local comparison. *Erdkunde, Vol. 73*(1), 19–29. https://doi.org/10.3112/erdkunde.2019.01.04

Griffiths, D., Sigona, N., & Zetter, R. (2005). *Refugee community organisations and dispersal: Networks, resources and social capital.* Policy Press.

Guiraudon, V., & Lahav., G. (2000). A Reappraisal of the State Sovereignty Debate. *Comparative Political Studies, 33*(2), 163–195. https://doi.org/10.1177/0010414000033002001

Jørgensen, M. B. (2012). The diverging logics of integration policy making at national and city level. *International Migration Review, 46*(1), 244–278. https://doi.org/10.1111/j.1747-7379.2012.00886.x

Kazepov, Y. (Ed). (2010). *Rescaling Social Policies towards Multilevel Governance in Europe*. Avebury.

Kazepov, Y., & Barberis, E. (2013). *Il welfare frammentato. Le articolazioni regionali delle politiche sociali italiane*. Bologna.

Keating, M. (1998). *The New Regionalism in Western Europe: Territorial Restructuring and.* Political Change. Northampton: Edward Elgar Publishing.

Kibreab, G. (2007). Why governments prefer spatially segregated settlement sites for urban refugees. *Refuge: Canada's Journal on Refugees, 24*(2007), 27–35. https://doi.org/10.25071/1920-7336.21365

Könönen, J. (2018). Differential inclusion of non-citizens in a universalistic welfare state. *Citizenship Studies, 22*(1), 53–69. https://doi.org/10.1080/13621025.2017.1380602

Kreichauf, R. (2018). From forced migration to forced arrival: The campization of refugee accommodation in European cities. *Comparative Migration Studies, 6*(1), 7. https://doi.org/10.1186/s40878-017-0069-8

Lebuhn, H. (2013). Local border practices and urban citizenship in Europe: Exploring urban borderlands. *City, 17*(1), 37–51. https://doi.org/10.1080/13604813.2012.734072

Lipsky, M. (2010). *Street-Level Bureaucracy*. Russell Sage Foundation.

Malkki, L. (1995). Refugees and Exile: From "Refugee Studies" to the National Order of Things. *Annual Review of Anthropology, 24*(1), 495–523.https://doi.org/10.1146/annurev.an.24.100195.002431

Marchetti, C. (2012). Oxford, 6–7 December. *Framing emergency. Italian response to 2011 (forced) migrations from Tunisia and Libya, Paper presented at the "Rsc 30th Anniversary Conference: Understanding Global Refugee Policy.*

Marchetti, C. (2014). Refugees and Forced Migrants in Italy: The Pendulum between 'Emergency' and 'System'. *REMHU: Revista Interdisciplinar Da Mobilidade Humana, 22*(43), 53–70. https://doi.org/10.1590/1980-85852503880004304

Marchetti, C. (2016). Le sfide dell'accoglienza. Passato e presente dei sistemi istituzionali di accoglienza per richiedenti asilo e rifugiati in Italia. *Meridiana*, 86.

Martinelli, F., Anttonen, A. H., & Maetze, M. (2017). Social Services disrupted: Changes, challenges and policy implications for Europe in Times of Austerity. Edward Elgar.

Marzorati, R., Semprebon, M., & Bonizzoni, P. (2017). Piccolo è bello? Governare l'immigrazione nei piccoli comuni lombardi. *Mondi Migranti, 1*(1), 113–130. https://doi.org/10.3280/MM2017-001007

Mayblin, L., & Poppy, J. (2019). Asylum and refugee support in the UK: Civil society filling the gaps? *Journal of Ethnic and Migration Studies, 45*(3), 375–294. https://doi.org/10.1080/1369183X.2018.1466695

Médecins sans Frontières (2016). *Fuori campo. Richiedenti asilo e rifugiati in Italia: insediamenti informali e marginalità sociale*. Roma: Médecins sans Frontières . *Retrived from* http://www.medicisenzafrontiere.it/notizie/news/fuori-campo-mappa-dell%E2%80%99accoglienza-che-esclude Last access: 10th Feb 2019

Mezzadra, S. (2011). The gaze of autonomy: Capitalism, migration and social struggles. In V. Squire (Ed.), *The Contested Politics of Mobility: Borderzones and Irregularity* (pp. 121–142). Routledge.

O' Mahony, L., F., & Sweeney, J., A. (2010). The exclusion of (Failed) Asylum Seekers from Housing and Home: Towards an Oppositional Discourse. *Journal of Law and Society, 37*(2), 285 314. https://doi.org/10.1111/j.1467-6478.2010.00505.x

Oomen, B., Davis, M., & Grigolo, M. (2016). *Global Urban Justice: The Rise of Human Rights Cities*. Cambridge University Press.

Pavolini, E. (2015). How many Italian welfare systems are there? In U. Ascoli (Ed.), *The Italian Welfare State in a European Perspective* (pp. 283–301). Bristol.
Pinelli, B. (2017). Control and Abandonment: The Power of Surveillance on Refugees in Italy, During and After the Mare Nostrum Operation. *Antipode*, *50*(3), 725–747. https://doi.org/10.1111/anti.12374
Pitzalis, S. (2018). La costruzione dell'emergenza. Aiuto, assistenza e controllo tra disastri e migrazioni forzate in Italia. *Argomenti*, *10*(2018), 103–132.
Purcell, M. (2006). Urban Democracy and the Local Trap. *Urban Studies*, *43*(11), 1921–1941. https://doi.org/10.1080/00420980600897826
Ranci, C., Brandsen, T., & Sabatinelli, S. (2014). Ranci, Brandsen, C. T., SabatinelliS, Local welfare systems in Europe in the age of austerity. In *Social vulnerability in European cities. The role of local welfare in times of crisis* (pp. 273–298). Palgrave: Macmillan.
Rauch, D. (2008). Central versus Local Service Regulation: Accounting for Diverging Old-age Care Developments in Sweden and Denmark, 1980–2000. *Social Policy & Administration*, *42*(3), 267–287. https://doi.org/10.1111/j.1467-9515.2007.00596.x
Rigo, E. (2011). Citizens despite Borders: Challenges to the Territorial Order of Europe. In Squire, V. (Ed.), *The Contested Politics of Mobility. Borderzones and Irregularity*, 199–215. Routledge. .
Robinson, V., Andersson, R., & Musterd, S. (2003). *Spreading the burden*. Bristol.
Sabatinelli, S., & Semprebon, M. (2017). Changes in the vertical division of labour and the rescaling of authority on social services within the state. Better tailored services or de-responsibilisation? In F. Martinelli, A. Anttonen, & M. Mätzke (Eds.), *Social services disrupted Implications and policy challenges for a crisis-ridden Europe* (pp. 114–133). Cheltenham: Edward Elgar.
Sainsbury, D. (2012). *Welfare states and immigrant rights: The politics of inclusion and exclusion*. Oxford University Press.
Sanò, G. (2018). *Spazio prodotto e Spazio produttore. La relazione tra il sistema di accoglienza e due territori a Messina, in Bergamaschi M. Piro V. (a cura di) Sociologia urbana e rurale* (pp. 59–76). Milano.
Scholten, P. (2011). *Framing Immigrant Integration: Dutch Research-Policy Dialogues in Comparative Perspective*. Amsterdam University Press.
Scholten, P. (2013). Agenda Dynamics and the Multi-Level Governance of Intractable Policy Controversies: The Case of Migrant Integration Policies in the Netherlands. *Policy Sciences*, *46*(3), 217–236. https://doi.org/10.1007/s11077-012-9170-x
Schuster, L. (2004). The exclusion of asylum seekers in Europe, *COMPAS Working paper*. Retrieved from https://www.compas.ox.ac.uk/media/WP-2004-001-Shuster_Asylum_Europe.pdf
Scotto, A. (2018). *Emergenza permanente L'Italia e le politiche per l'immigrazione*. Epoké: Novi Ligure.
Semprebon, M. (2017). Il protagonismo dei rifugiati. Frammenti di inclusione lavorativa in. Valle Camonica & M. Colombo Eds., *Immigrazione e contesti locali*. Annuario CIRMiB (305–315). Vita e Pensiero.
Semprebon, M., & Pelacani, G. (2020). Dispersal and reception in Northern Italy: Comparing systems along the Brenner route. In B. Glorious & J. Doomernik (Eds.), *Geographies of Asylum in Europe and the Role of European Localities* (pp. 15–44). IMISCOE Book Series.
Semprebon, M., & Caroselli, S. (forthcoming). Seekers and holders of international protection in Bozen: Arrival, transit and reception within an internal border area. In F. Della Puppa & G. Sanò (Eds.), *Across the boundaries of reception: Social trajectories, material conditions and coping strategies of refugees and asylum seekers inside and outside the Italian System of International Protection*. Università Cà Foscari.
Semprebon, M., Caroselli, S., & Scarabello, S. *(forthcoming). Coping with the evolving trends of human trafficking in Italy before and during the COVID-19 pandemic*. Università IUAV di Venezia.
Shutes, I. (2016). Work-related Conditionality and the Access to Social Benefits of National Citizens, EU and Non-EU Citizens. *Journal of Social Policy*, *45*(4), 691–707. https://doi.org/10.1017/S0047279416000234

Sorgoni, B (2011). *Etnografia dell'accoglienza. Rifugiati e richiedenti asilo a Ravannea.* Cisu: Ravenna.

Soysal, Y. (1994). *Limits of citizenship: Migrants and postnational membership in Europe.* University of Chicago Press.

Spencer, S. (2018). Multi-level Governance of an Intractable Policy Problem: Migrants with Irregular Status in Europe. *Journal of Ethnic and Migration Studies, 44*(12), 2034–2052. https://doi.org/10.1080/1369183X.2017.1341708

Squire, V. (2009). *The Exclusionary Politics of Asylum.* Palgrave Macmillan.

SSIIM Unesco Chair. (2018a). *I servizi rivolti a (o con alto tasso di) utenti immigrati nella Città Metropolitana di Venezia.* REPORT sulle CRITICITÀ E POTENZIALITÀ. Available at: http://www.unescochair-iuav.it/capacitymetro/documenti/. Last access: 9 Nov. 2020.

Vermeulen, F., & Stotijn, R. (2010). Local policies concerning unemployment among immigrant youth in Amsterdam and in Berlin: Towards strategic replacement and pragmatic accommodation." In T. Caponio & M. Borkert (Eds.), The local dimension of migration policymaking (pp. 109–134). Amsterdam University Press.

SSIIM Unesco Chair. (2018). I servizi rivolti a (o con alto tasso di) utenti immigrati nella Città Metropolitana di Venezia. *REPORT sulle RETI TERRITORIALI.* Italy: University IUAV of Venice. Available at: http://www.unescochairiuav.it/capacitymetro/documenti/. Last access: 9 Nov 2020.

Van Der Leun, J., & Bouter, H. (2015). Gimme Shelter: Inclusion and Exclusion of Irregular Immigrants in Dutch Civil Society. *Journal of Immigrant & Refugee Studies, 13*(2), 135–155. https://doi.org/10.1080/15562948.2015.1033507

Varsanyi, M. W. (2006). Interrogating "Urban Citizenship" vis-à-vis Undocumented Migration. *Citizenship Studies, 10*(2), 229–249. https://doi.org/10.1080/13621020600633168

Vianelli, L. (2014). *Frustrazione/Potenzialità. Il sapere antropologico nella quotidianità di un progetto di accoglienza i rifugiati e richiedenti asilo.* A.L. Palmisano. a cura di). Antropologia applicata.

Vosko, L. F., Preston, V., & Latham, R. (2014). *Liberating Temporariness? Migration, Work and Citizenship in an Age of Insecurity. McG.* ill-Queen's University Press.

Whyte, Z. (2011). Enter the myopticon: Uncertain surveillance in the Danish asylum system (Respond to this article at http://www.therai.org.uk/at/debate). *Anthropology Today, 27*(3), 18–21. https://doi.org/10.1111/j.1467-8322.2011.00808.x

Zetter, R. (2007). More Labels, Fewer Refugees: Remaking the Refugee Label in an Era of Globalization. *Journal of Refugee Studies, Vol. 20*(2), 172–192. https://doi.org/10.1093/jrs/fem011

# Making urban humanitarian policy: the "neighbourhood approach" in Lebanon

Romola Sanyal

**ABSTRACT**
Displaced people are increasingly living in urban areas and humanitarian organisations are rethinking their policies and practices. The 'Neighbourhood Approach'-an area-based policy model has become globally popular amongst humanitarians. In this paper, I trace its development in Lebanon through a Temporary Technical Committee (TTC) on Neighbourhood Upgrading. Although it failed in being taken up as a distinct policy, aspects of it are being incorporated into plans and agendas of various actors. Through a critical document analysis I interrogate how humanitarians imagine ideas of 'neighbourhood' and 'community' in urban contexts. Using the critical literature on urban policymaking and mobilities, I show how the making of the neighbourhood approach draws together people, experiences, lessons, and territories both near and far, thus complicating its provenance as local or global. I offer a glimpse into the world of urban policymaking by humanitarian organisations whilst also challenging the mainstream discussions on urban policy mobilities.

## Introduction

The Syrian conflict which began in 2011 has displaced millions of people into neighbouring countries such as Turkey, Jordan and Lebanon. Some have travelled further on to Europe and other parts of the world in search of safety. In countries such as Jordan and Turkey, refugee camps have been set up for displaced Syrians. In others such as Lebanon, there has been a ban on refugee camps, and Syrians have either found shelter in cities, or in informal tented settlements (Bergby, 2019; Fawaz et al., 2014; Sanyal 2017). Regardless of having camps, large numbers of displaced Syrians have moved to urban areas in host countries in the Middle East, often into poor areas, where they can access cheaper housing options. In many instances this has led to increased crowding in neighbourhoods as buildings and apartments are modified and divided to accommodate more people (Boustani et al., 2016; Fawaz, 2016). Billions of dollars in aid, international organizations and their experts have also arrived to "manage" the crisis. Local and global actors have brought varying experiences, pedagogical training and expertise to bear on the situation that is simultaneously global yet ineluctably local in how it evolves. With the protraction of the crisis, new methods of humanitarian interventions are being tested, drawing on lessons from further afield, but also creating ideas from the local context.

In this paper, I analyze the brief development of one such policy- the Neighbourhood Approach- in Lebanon. Although a glimpse, I show how this global policy came to be developed through experiments and negotiations that organizations and NGOs were undertaking locally. In particular, through a careful document analysis, I interrogate the ways in which ideas of 'neighbourhood' and 'community' were imagined by contributors to the text. Here, I draw on the critical geographical literature on policy formation and mobilities to analyze how a policy is arrived at, and how it is an impossible entanglement of the local and the global. In doing so, the paper makes two contributions – firstly to show how humanitarians 'learn' urban work and how their interventions can challenge how urban territories are conceptualized and acted upon. In relation to that, the second is to shift away from the focus on urban entrepreneurialism in policy mobilities literature by drawing attention to the ways cities in zones of 'crises' become models for policies as well, especially as humanitarians become key actors in shaping and mobilising urban policies and urban futures.

## Geographies of policy

The project I initiated in Lebanon in 2014 was to understand how humanitarian organizations were responding to an 'urbanised crisis' and what challenges their interventions raised in the contexts where they operated. As part of the project, I, together with my research assistant interviewed numerous NGOs both local and global in different parts of the country. In the process, we stumbled across the development of the 'Neighbourhood Approach' which was being carried out by a few international NGOs (INGOs) but also being comprehensively developed by UN Habitat together with several INGOs[1] through a Temporary Technical Committee (TTC). The intention was to brainstorm and develop a single policy toolkit to improve urban humanitarian responses in Lebanon. Though several meetings took place in 2016 and drafts of guidelines were developed through discussions and input by various organizations, particularly operating in the shelter sector, the 'Neighbourhood Approach' as a discrete policy failed to materialise as it was originally envisioned due to various factors including objections by the government. Elements of it have been continued by various multilateral organisations, several NGOs have also continued to pursue it within their own organisations, and urban responses, multisectoral area-based approaches have appeared in later crisis response plans and have broadly been supported by the Lebanese government.

The Neighbourhood Approach as devised within the Lebanese context drove me to study the broader and emerging landscape of urban humanitarian policy-making. Through it, I traced the ways in which area-based approaches (one of several terms referring to a geographically fixed approach) have been circulating across the world since 2010. The scholarship on urban policy-making is particularly salient here. As several academics argue, we have moved away from a problematic political science analysis of policy transfer to a geographical analysis of policy mobilities where attention is given to the active construction of networks of policy circulations. This literature is also attentive to how policies mutate as they move and materialise in different territories (McCann, 2011; Peck & Theodore, 2010; Prince, 2017; Robinson, 2015). The mobility of policies offers a lens through which we study the relational nature of cities and their politics, how they are assembled and linked to and by people, ideas, and objects near and far (ibid).

However, this approach has been rightly critiqued for attending largely to tracing the movement of policies and paying less attention to how policies are 'arrived at' in the local context (Robinson, 2013, 2015). In doing so, inadvertently, the policy-mobilities literature contributes to a global/local dualism and exacerbating presumed differences between what comes from out there and what happens here (Prince, 2017). Essentially, it becomes difficult to know if a policy adoption is the result of local actors drawing on globally circulating policies or if it is the result of local specificities (Prince, 2017; Robinson, 2015). To overcome this, scholars have urged a shift away from topographical understanding of policy development to a topological one (ibid). In other words, to understand how policy makers compose certain policies instead of others, it is useful to shift away from tracing the mobilities of policies to turn to the local context and consider how the making of policies involves the borrowing and invention of ideas simultaneously (Robinson, 2015). Drawing on her work in Johannesburg, Robinson argues that how we may want to start is by asking *why those policies*? In posing this question, she shifts our focus away from tracing the arrival of the policies to thinking about the "incredibly messy, often untraceable processes of policy formulation" (Robinson, 2013, p.11). We consider how diverse groups of people draw on different ideas from distinct sites, temporal moments, documents, influences from near and far to come up with policies that are specific and unique to their own context. She notes that ideas may have always already been in place, or they have arrived in different ways, or not arrived at all, or forgotten, or put away. This returns us to the specifics of the local, how it is made of up multiple elsewheres, of different temporalities and materials that come together in unexpected and unpredictable ways, such that separating out what is 'global' and what is 'local' what is 'borrowed' and what is 'invented' becomes impossible, and even less relevant.

Robinson's provocation of shifting our starting point to asking how policies are 'arrived at', paying attention to how different elsewheres figure, together with present exigencies is instructive for this paper. The Neighbourhood Approach may have been a 'global policy' and may have 'landed' in Beirut through the circulation of experts, but the specific formulation of this policy is equally made up of other territories, people and knowledge. It borrows from the northern city of Tripoli, from the suburbs of Beirut, and neighbourhoods Mount Lebanon. It is also shaped through other histories of the country-of having hosted refugees for generations, of itself being the subject of humanitarian development especially after the 2006 conflict between Israel and Hezbollah. The policy then is that of multiple elsewheres and times brought together through what is written and omitted. Drawing further on Robinson (2018), and expanding beyond the local context, what is also important to consider is how the urbanisation of crisis management and humanitarian action offers new imaginations of urban concepts-in this case 'the neighbourhood' and 'community' and new ways of governing them.

The data from this project are necessarily partial, drawing on interviews with several organisations that were experimenting with the Neighbourhood Approach, as well as following the developments of the TTC between March and June 2016. The TTC was formed from the Shelter working group specifically to develop the approach/guidelines. It was comprised of a handful of INGOs and led by UN-Habitat. I engage in a close document analysis of working drafts of guidelines developed by them that came via e-mail exchanges. These were rich in commentary and edits by various members of the committee, which offered a unique insight into the process of policymaking. The document has a number of different discussions, but I focus on two specific issues, around defining neighbourhoods and communities, and how these evolved through input by various committee members over the course of these months. Further, my research assistant also attended four of the technical committee meetings which were semi-public. I use these observations to provide a contextual setting for the discussions. Where the notes comes from her, this is specifically cited in the endnotes. Meetings and negotiations between different organisations continued well past when this project ended, but I do not have this material. However, I note that I separately continued to follow the work of two of these organisations from 2017 until 2020 through key informant interviews and studying internal organisational documents. I supplement this with analyses of documents on neighbourhood approaches and area-based approaches created by organizations working in Lebanon and elsewhere in the world.

## Urban displacement, planning and humanitarian expertise

Despite long histories of displaced people living in cities in different parts of the world there has been a tendency in academic and policy literature to focus on camps rather than urban environments. Arguably an 'urban turn' in refugee studies has been brought about by a belated acknowledgment and acceptance by the United Nations High Commissioner for Refugees (UNHCR) in 2009 of urban displacement (Darling, 2017). While scholars have expanded the scope of work on urbanised displaced populations, several disciplines such as urban planning, which would seem centrally interested in studying urban refugees have continued to be largely disengaged from these discussions. Part of the reason why displaced populations do not yet fall into mainstream planning literature may be due to the fact that they are still seen as a humanitarian concern, and as temporary inhabitants in cities. It may also be because attending to the needs of refugees may not be of interest to local authorities and planners, or because helping displaced people may be controversial and counterproductive, particularly in places where local populations are antagonistic toward displaced people (Kihato & Landau, 2016). Yet, displaced populations can and should be included in urban analyses and policies because they too are city-makers and have rights to the city (Sanyal, 2014).

The engagement with urban issues in humanitarian practice is also relatively new. While there has been an expanding academic discourse on blurring the boundaries between camp and city, in humanitarian practice, this divide has been significant and persistent. Humanitarian aid has been rooted in working in camps and rural areas and is largely driven by the prevailing cluster approach and sector-based delivery (Bergby, 2019; Campbell, 2018; Sanderson, 2019).[2] This approach is ill-suited for urban environments that have preexisting institutional structures and systems and are complex, contradictory, and dynamic. In order to 'urbanise' their approaches, many organizations are finding it necessary to retool their approaches away from traditional models, but have found this to be extremely challenging. They now have to be agile and flexible, work with a variety of stakeholders including municipal governments, private sector actors and to coax donors to shift away from the traditional aid architecture. They also need to move toward inter-cluster, multi-sectoral responses (Sanderson, 2019, 2020). This shift is difficult to undertake as crises proliferate and become increasingly protracted. One could perhaps argue that the city has become a different kind of camp in which experiments in service provision and governance are being undertaken. In the Lebanese context, this metaphor can be pushed further as organizations 'carve up' different neighborhoods in which to work, as will be discussed further on. In a way then, neighbourhoods come to take the place of camps as sites of humanitarian work.

The 'urban turn' in humanitarianism has also brought about the creation of urban toolkits. The earthquake in Haiti appears as a key 'origin story' in the development of urban responses for humanitarian action. It is after this disaster that various aid organizations 'woke up' to the need to urbanise their responses (Sanderson, 2019). But, this was not the only 'site' where the urbanisation of humanitarianism took place. Cities such as Kabul (Afghanistan), Bangui (Central African Republic), Tacloban and Bogo (Philippines) amongst others have all been sites of 'urban learning' for humanitarian actors.[3] The Syrian Crisis has also provided an important spatio-temporal opportunity for changing and advancing humanitarian practices. Here too, humanitarian organizations are shifting from a humanitarian and emergency focus to one that increasingly identifies resilience, sustainability and self-reliance (Gabiam, 2016) and often in urban areas. The neighbourhood approach is a key tool in the development of urban responses in the Middle East, in countries such as Jordan and Lebanon, but policies 'territorializing' the neighbourhood and community illuminate how these ideas come to be formed through local specificities. In the next section, I briefly discuss the role of neighbourhoods in planning, and in Lebanon, to contextualise the broader argument.

## The neighbourhood as a concept

The neighbourhood has long been a key site of socio-spatial relations and social control in human societies and settlements (Abraham, 2018; Kallus & Law-Yone, 2000). In Western planning, as Kallus and Law-Yone (2000) note, the self-conscious development of the idea as a key part of urban design and governance is a more modern phenomenon, and it is also arguably more significant in this context than in the Global South. Yet, even here, what the neighbourhood is precisely is subject to considerable debate. Like its related term 'community', 'neighbourhood' is an invented and deeply contested concept. As many scholars have noted, neighbourhoods may mean different things to different people and each resident can bring their own understanding of neighbourhood boundaries (Blokland, 2009; Stein, 2014). Here, race and class and other identities play crucial roles in how neighbourhoods and their boundaries are imagined and mobilised. Therefore, it is important to be attentive to who are seen to be 'residents' of a neighbourhood and authorised to speak on its behalf (Blokland, 2009).

The neighbourhood has also been an important part of decentralisation processes in the Global North, and this has viewed with considerable cynicism. Academics have been critical of the problematic imposition of boundaries, the assumptions around 'community' and democratic legitimacy at this scale (Cowie & Davoudi, 2015) and more fundamentally, critiqued the turn toward the local scale for masking the workings of neoliberalism under the guise of democracy (Brenner & Theodore, 2002; Parker & Street, 2015; Lord & Tewdwr-Jones, 2014) . Related to these are concerns about the means and methods of participation in neighbourhood planning. Fundamentally, there is an implicit assumption that neighbourhoods are homogenous, that there are shared values and interests, and that these would then translate into consensus over plans. However, as scholars have pointed out, this is problematic as the question of consensus is contentious at best. Discussions and dialogue are imbued with power politics in which some voices are silenced by those who are more powerful. Furthermore, although participation in planning issues is encouraged, it is also resource intensive, and inadvertently a smaller group of community representatives are involved – often the usual suspects, skewing consultations and making them exclusionary. Therefore, valid questions can be raised about who would be participating and who would be left out, how to build social capital across diverse groups affected by plans and projects, and ensure that those groups with already strong social capital do not act in insular and exclusionary ways (Holman & Rydin, 2013). Thus, although viewed as democratic, identifying neighbourhoods, neighbourhood boundaries, and engaging in planning practices that mobilize neighbourhoods and communities are fraught processes that rest on some shaky assumptions about social relations between people, political agendas and invented boundaries.

The neighbourhood in the Lebanese context plays an important role in the socio-spatial transformations. Whilst a comprehensive discussion is not possible here, a few key points about the salience of neighbourhoods are important to highlight for this discussion. As (Fawaz, 2009) notes in her work, the production of neighbourhoods have been key for communities to place a foothold in cities such as Beirut and to exercise their right to the city in the post-independence era. Social networks and the entanglement of the formal and informal have remained crucial to the production of neighbourhoods, including informal ones (Fawaz, 2008). The most significant transformations to neighborhoods in Lebanon have occurred the civil war when massive population displacements took place, dividing cities like Beirut into homogenized, sectarian neighbourhoods. Wartime geographies continue to haunt the city, influencing where people live, travel and socialize (Deeb & Harb, 2013). The territorialization of sectarianism and the anticipation of future conflicts significantly affect the ways in which neighborhoods come to be formed and the kinds of services and infrastructures they can access. (Bou Akar, 2018; Nucho, 2016; Seidman, 2012) Along with sectarianism, neoliberal developments have also led to considerable changes as rising property prices have pushed out many residents from centrally located neighbourhoods in Beirut (Khechen, 2018; Sawalha, 2010). Finally, as will be noted further on in the paper, there is a dearth of data, particularly on poorer neighbourhoods and the kinds of resources and deprivations they face. All this makes for planning and interventions in neighbourhoods highly complex and politically fraught processes. Placing planning critiques together with a local history of neighbourhoods is important as we unpack the politics of 'neighbourhood', 'community' and 'participation' within humanitarian work in Lebanon.

## Area-based/settlements/neighbourhood-based approaches in humanitarian work

The neighbourhood scale is popular amongst humanitarian organizations and seen as the appropriate scale for attempting to "integrate" refugees and local residents (Seethaler-Wari, 2018). Neighbourhood approaches, also referred to as settlements-based approaches or area-based approaches (ABAs) are seen to enable more holistic responses to protracted and increasingly urbanised crises (Bergby, 2019). They have gained currency since their major advent during the post-Haiti recovery and have been popularized by USAID and increasingly supported by other organisations. Despite their 'origin story' being embedded within the Haitian recovery, scholars argue that their pedagogical foundations can be drawn back to the 1960s and 1970s to work undertaken by urban and regional planners in vulnerable neighbourhoods (Schell et al., 2020). Interestingly, however, this work refers to a development planning work undertaken in many countries in the Global South seen to be in 'need of interventions'. Further, it problematically collapses a number of different practices- urban re-generation, site-and-services, slum upgrading, integrated development programs- all of which have very different geographic and developmental trajectories into a single framework (Sanderson, 2017).

Area-based approaches rely on defined areas with high needs, and focusing on communities living in them. The approach is intended to be multi-sectoral- addressing not just shelter and WASH needs, but also considering health, livelihoods, protection and so forth, and intended to be participatory and inclusive (Sanderson, 2017; Elizabeth Parker & Maynard, 2015; Schell et al., 2020). Several questions can be raised here about how an area is defined, who is involved in the definition, how areas are divided between different organisations, how coordination is set up, and what the implications of these processes are on urban environments. More generally, the policy literature on area-based approaches focuses on the implications on humanitarian practice rather than considering, with any nuance, the implications on urban development, politics, or governance in the future. This forms part of a larger critique of humanitarian policy-making which is ephemeral, and subject to temporal priorities , donors, and exigencies occurring elsewhere.

Indeed, significant critiques have been raised about area-based approaches in humanitarian governance. (Parker & Maynard, 2015, p. 4) for example, note that an area-based approach is implemented in a number of different ways, using a range of terminologies and "as a result on-going humanitarian policy and operational discussions are not informed by a shared understanding of what 'area-based programming' means in practice, why, when or how to adopt the approach, nor the institutional implications, such as funding and administrative mechanisms." Expanding on that, they note that whilst there is some general commonality on focusing on the neighbourhood scale, targeting families ranging from 1400 to 6000 people, the scale of operations remains unclear. Thus, projects can focus on a few households to encompassing entire urban areas (ibid). This critique highlights the vagaries embedded in urban humanitarian policy making and its mobility.

## The Lebanese context: from emergency to stabilization

The specificities of the Lebanese context are fundamental to the evolution of the neighbourhood approach both within the TTC, and other organisations. Lebanon is not a signatory to the Refugee Convention, hence, Syrian refugees are not classified as 'refugees' but as 'displaced persons' or 'de facto refugees' which has no legal bearing and effectively renders Syrians in the country as migrants. This classification coupled with policies such as border restrictions, costly annual visa renewals, along with a host of other ad-hoc policies designed to deter Syrians have driven many into conditions of destitution and invisibility (NRC ICLA, 2014; World Bank, 2016; Sanyal 2018).

In 2014 the number of Syrians displaced into Lebanon surpassed 1 million. The government, was at the time, running without a president or parliamentary elections, and in effect had a 'caretaker government'. In response to the crisis, it followed a policy of inaction (Harb et al., 2018) whilst also denying UNHCR the possibility of setting up refugee camps. The Syrian population thus scattered across the country in rural and urban areas and attempted to acquire whatever shelter they could access. The influx of large numbers of refugees into many municipalities has overwhelmed local communities and compounded the already existing woes of local administrators. It should be noted here that Lebanon is no stranger to urban refugees: Armenian, Palestinian, Iraqi, and other refugees have found shelter in the country for a very long time. Palestinian camps which were once on the outskirts of cities such as Beirut have become part of the urban fabric thus blurring the boundaries between cities and camps (Martin, 2015; Sanyal, 2014).

The influx of large numbers of refugees into Lebanon has had a significant impact on local governance particularly as municipalities have been on the frontline of hosting refugees, often against a backdrop of limited or no support from the central government (Boustani et al., 2016). Lebanon has a largely centralised governance system despite being committed to decentralisation. In a four-tiered governance structure, the municipality, is the smallest and the only autonomous and elected body (Harb & Atallah, 2015). There are a very large number of municipalities, many of which are very small, with limited financial or administrative means (Atallah, 2016). This impacts their ability to do more holistic planning, relying instead on more sporadic interventions and on international aid, particularly in responding to the Syrian crisis (Bergby, 2019). Some of these have been grouped into Unions of Municipalities and coordination between them in certain parts of the country work well, making humanitarian efforts there also easier.[4] Furthermore, there is a lack of data on the urban environment, services, population, and needs, which affects the possibilities of undertaking planning interventions in the country (ibid). Finally, and importantly, as noted earlier, Lebanon's government is structured on a confessionalist system and the sectarian logic permeates all levels of governance, planning institutions, and socio-spatial transformations. Bou Akar (2018, p. 147) notes that in Lebanon "..urban planning has devolved into a series of practices that produce the spatiality of sectarian difference while facilitating continued, profitable real estate development." The Syrian war has not just impacted the population increase within Lebanon, but has affecting political and sectarian politics and divisions in the country causing conflicts and solidarities to become further entrenched. The combination of contested planning, limited means, and sectarian urbanisation fundamentally affects the ways in which humanitarian policies come to be formed in the country.

Humanitarian organisations that had either been working in the country previously or arrived due to the onset of the crises began undertaking a standardized, individualized responses – targeting only Syrian families with emergency supplies and services. This proved to be deeply problematic as Lebanese host families and neighborhoods who had faced decades of poverty and deprivation, and yet were extending their hospitality were being bypassed.[5] They were thus doubly abandoned – first by their own government, and then by aid agencies. As the crisis became more protracted and the burden of hospitality became more difficult to shoulder for many host communities, inter-communal tensions began flaring. Aid agencies began rethinking many of their strategies and considered more comprehensive approaches to assistance that would serve both refugees and the local host communities. The Neighbourhood Approach, which was already a global policy, became increasingly relevant in this context. Aspects of it were included in the Lebanon Crisis Response Plan (LCRP) of 2015/16 and later. The LCRPs urged a rethinking of humanitarian aid, to include both Syrians and Lebanese families and communities, especially the poor who were hosts. In doing so, the plan suggested a shift from emergency relief to stabilisation and the involvement of the government, particularly through the Ministry of Social Affairs (Government of Lebanon, 2017; Government of Lebanon and United Nations, 2014).

## Moving to a neighbourhood approach

In Lebanon, spaces in urban and peri-urban areas expanded to accommodate the large numbers of Syrian refugees. However, the resultant increased levels of overcrowding and densification affected the quality of housing and living conditions in many neighbourhoods including informal ones (Fawaz, 2016).

To urbanise their responses, many NGOs began experimenting with neighbourhood approaches[6] anticipating that a geographical approach supporting both Syrians and Lebanese, would reduce conflicts between them and also be cost efficient. It could also potentially improve relations between organisations, local communities and governments, in particular, where local governments were brought on board and consented to humanitarian operations. Some undertook assessments in different parts of the country such as Mount Lebanon to determine where gaps lay and where their interventions could be most effective. They then used that to carve out distinct spaces in which they alone would operate.[7] Others who had established their presence in certain cities, such as Tripoli, worked together with LNGOs to assess the most vulnerable neighbourhoods and their needs and base their funding on that. This cemented closer ties between INGOs and LNGOs that was beneficial for local capacity building but also for finer-grained assessments and understandings of local areas and communities[8] (Boustani et al., 2016). In 2016, most organisations we encountered were either in the assessment stage or at the very early stages of working out their neighbourhood level projects. Their experiences and expertise in terms of community engagement, local government engagement, doing assessments in different parts of the country informed the comments they made on drafts of guidelines for the Neighbourhood Approach.

In 2015, the TTC on Neighbourhood Upgrading began developing guidelines for the policy. The documents noted that the terminology used was based on global guidelines on Neighbourhood and Area-Based Approaches, however the ideas were clearly driven by urban lessons learned from Lebanon. The document contained distinct ideas around Neighbourhood Approach and Neighbourhood Upgrading. These two ideas overlapped to some extent, but were also distinct. According to the draft document, the Neighbourhood Approach was "enhanced humanitarian assistance mainly in urban areas, through basing interventions on neighbourhoods, providing coherent and multi-sectorial assistance amidst the multi-faceted conditions of urban areas. Thus, a 'Neighbourhood Approach' was an area-based means of responding to multi-sector needs that is informed by a community-based decision-making process reflective of the social, economic, and physical features of the delimited area." Neighbourhood

Upgrading on the other hand involved a "formula taking into account space, population, shelter vulnerability, infrastructure and service delivery needs, utilization of existing resources, and a minimum package of interventions. These included: shelter, water and sanitation upgrades, rehabilitation of buildings common facilities, maintenance, upgrades, and/or extension of existing basic urban services infrastructure, general and environmental improvements of common spaces/public areas and community management and maintenance". In other words, the " '*Neighbourhood Approach*' described the overarching approach [which included 'soft components' such as protection and enhancing social services] while the '*Neighbourhood Upgrading*' described the intended outcomes of this activity-often related to shelter and WASH, and tools to achieve these." That the document was titled Neighbourhood Upgrading, rather than Approach is telling, as it signals the central role of the shelter sector in formulating and implementing these guidelines. In addition, the guidelines having an upgrading focus had a pragmatic angle- some agencies felt that they did not have the capacity to take on a full neighbourhood approach that would entail holistic assessments and strategies for interventions across all sectors. Rather, they wanted to focus on physical surroundings, ideally in partnership with others working on more 'social' issues. There were some discussions that such a split could possibly be problematic as a focus on physical improvements, while useful, may overlook far more pressing needs in an area and that demand greater attention.[9]

The TTC attempted to meet on a weekly basis initially to develop a series of guidelines for organizations involved in this work. However, over the course of three months some were postponed or canceled due to various constraints. Attendance at meetings also fluctuated between having anywhere from nine to three participants from different organisations.[10] Drafts were circulated over e-mail with comments from various committee members. Participants brought their own organisational, work experience, and cultural backgrounds to bear on how they input into the changing document.

Although the initiative to this response sat with the Shelter working group, several sectors contributed to the strategy, particularly the Water and Energy, Protection and Social Cohesion sectors. Later drafts dropped the language of having the shelter group leading this committee and the approach being settlement-focused to make it more inclusive and draw attention to the social dimensions of neighborhoods in addition to the physical and spatial aspects of them. However, the approach continues to be dominated and driven by shelter priorities today.[11]

## Defining and bounding a neighbourhood

The TTC also spent a considerable amount of time, particularly in the initial period debating how to define and demarcate neighbourhoods. For these organisations, it was a fine balance between understanding social relations on the ground to improve their service delivery, and having a practicable scale to carry out their work. In other words, the size of the neighbourhood was influenced by the pragmatics of humanitarian project work including budgetary and capacity constraints, donor expectations, and priorities of organisations, as well as local specificities of Lebanon. The working group relied on the definition of neighbourhood based on that used by USAID. This description of neighbourhood, which itself appears to rely on western logics of what a neighbourhood is and does was as follows: "*Neighbourhoods are geographic areas of cities typically defined by social, economic, and physical features, which often serve as the basis for administrative and political recognition within larger jurisdictions. Neighbourhoods provide their residents with an identity and foothold in the larger urban area.*"

In early versions of the document, it was clear that defining and demarcating neighbourhoods was very much a product of the specificities of Lebanese urbanisations alongside the unique humanitarian situation unfolding in relation to that. There was an acknowledgment that neighbourhoods could be defined through boundaries and features involving administrative (existing administrative areas) and physical (natural barriers, roads, rivers or topography) criteria. It was also clear that municipal definitions of neighbourhoods were not satisfactory. Much like in planning, this is because such demarcations of neighbourhoods may not match the perceptions of neighborhoods or sub-neighbourhoods imagined by the resident communities themselves. In larger urban areas, boundaries between different neighbourhoods may also be fuzzier and interventions in one sub-area may potentially lead to tensions in another. Considerable time was thus spent on defining a neighbourhood that would be socially meaningful but operationally viable producing a Neighbourhood Approach that was uniquely local. For the purposes of upgrading, neighbourhoods were defined through shared needs, based on similar experiences, backgrounds and so forth. Members suggested using administrative and physical boundaries together with social analysis and delineation to create suggestive and provisional boundaries in a draft circulated in March. These boundaries would not only be fuzzy, but be subject to amendment based on work experience and budgetary constraints to avoid tensions between different areas.

The need for having boundaries was important for coordination between partners and municipalities. Most NGOs preferred working in specific areas so it was important to ensure that operations did not overlap. There was also an agreement that sometimes a neighbourhood would be larger than the capacities of a single NGO and would thus be divided into operational sub-areas.[12] Organisations were also expected to update systems to indicate when and what works were completed. However, a key factor in determining neighbourhood boundaries was a scale that was workable for NGOs. There were some debates in a later draft circulated in April about whether size or population should be included in the definition of a neighbourhood- whether it should be 0.005 sq km, or, as per the interventions of certain NGOs, at the scale of 5000 people. Some commented that where they came from, such a size/density would be considered a town. Others noted that socio-economic criteria should be considered when outlining a neighbourhood, rather than focusing on geographical boundaries.

Further, significant conversations took place around questions of informality and illegality. Again, different Lebanese territories folded into the analysis of these issues. Many organisations faced these issues especially in neighbourhoods where needs were high, that were either illegal, informal, or both. Informality also operated at different geographical scales – as entire neighborhoods, or as parts of buildings, which were constructed without permission, creating a patchwork of legality and illegality.[13]

From the March drafts, there were concerns raised about how there were conflicts of interest for ministries and municipalities regarding informal neighbourhoods where many services were not provided (Ministry of Water and Energy acknowledged that infrastructure plans did not include informal neighbourhoods). The drafts noted that upgrading would not be considered there. Some organisations working in informal neighbourhoods in Beirut, noted that they had been forced to remove things they had implemented with the permission of the municipality.[14] Discussions therefore commenced about how to intervene in them formally, and what level of government to involve in conversations around them- the municipalities from whom endorsements for working in these neighborhoods may be difficult, or the national government, or specific ministries involved in the maintenance of infrastructure, or security forces that need to provide approval for intervening in certain 'illegal' areas. The concerns revealed how the complex, overlapping and contradictory nature of governance in Lebanon often made coordination between humanitarian organizations and the state cumbersome and ineffective and lead to mistrust between them (Boustani et al., 2016) .

The last draft that I was able to access in June 2016 was a considerably evolved document, with hard and soft components that would form the core components of the Neighbourhood Approach (see Figure 1):

| | Components | | |
|---|---|---|---|
| | Hard | Soft | |
| **Dimensions** | Upgrading of Units | Stakeholder engagement | Municipality<br>Neighborhood committee |
| | Networks improvement, public/street space | Exit strategy | |

| | **Components** | | |
|---|---|---|---|
| | **Hard** | **Soft** | |
| **Dimensions** | Upgrading of Units | Stakeholder engagement | Municipality |
| | | | Neighbourhood committee |
| | Networks improvement, public/street space | Exit strategy | |

**Figure 1.** Table from draft document of the Temporary Technical Committee. Courtesy of TTC on Neighbourhood Upgrading, June 2016.

While neighbourhood definition was seen to be flexible, a suggestive geographical size was provided based on experiences of partners already undertaking this work. It was as follows:

> "Based on the principles of identification of neighbourhoods and the examples of neighbourhood sizes, the definition of neighbourhoods for the purpose of neighbourhood upgrading could therefore rely on three main parameters: Size; an average of 0.05sq km/5 hectares, Population/density, Budget (per capita?), and may be further categorized according to additional parameters, e.g.; socio-economic vulnerabilities, formal vs informal residential typologies and basic urban and social services. This would be defined through neighbourhood assessment/profiling … .The formula gives priority to the critical requirement to communicating budget needs to donors and partners to meet needs by applying the neighbourhood upgrading."

It noted that neighbourhoods could be identified through administrative and physical boundaries combined with social analysis and delineation. Larger neighbourhoods may also be subdivided according to such community structures, though keeping in mind interventions in one sub-neighbourhood may be challenged by adjacent sub-neighbourhoods with equal needs (see Figure 2 for methodology).

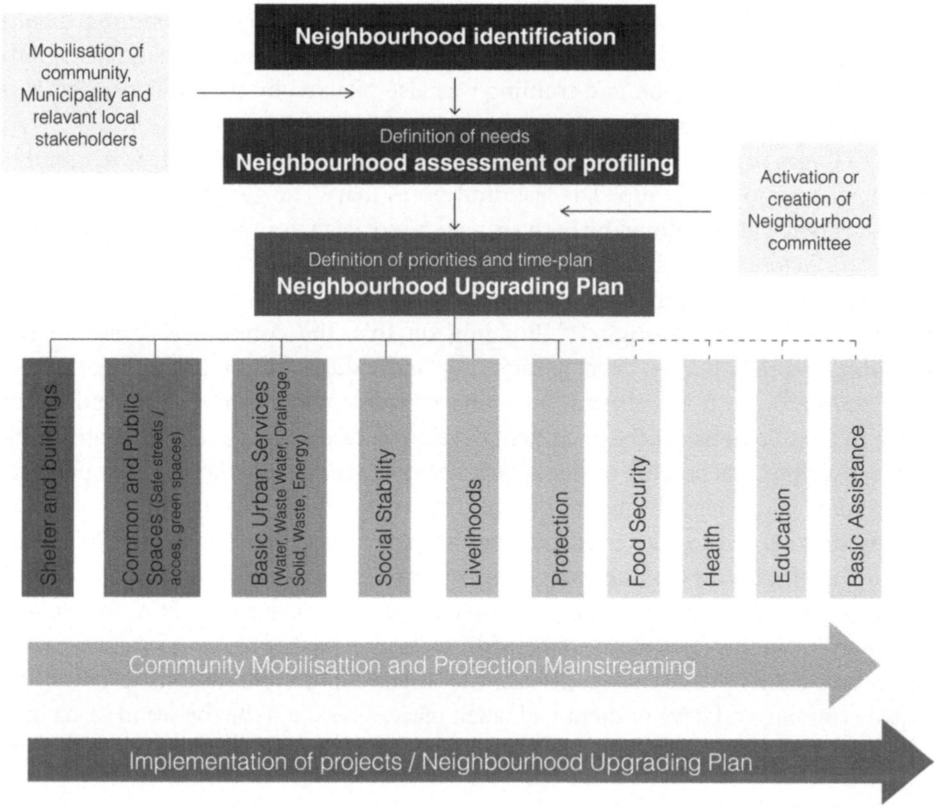


**Figure 2.** Diagram of methodology. courtesy of TTC on Neighbourhood Upgrading, june 2016.

## Involving communities

Communities and committees were seen to be key actors in the understanding of neighbourhood boundaries and the implementation of projects. In the charged, sectarian context of Lebanon, where attitudes toward refugees have been shifting, finding, constructing, and mobilising communities across Syrians and Lebanese was a complex task. In the NGO working in Tripoli, neighbourhood approaches were done through committees formed from the communities themselves. These committees made up of 6–7 people were mostly women despite attempts at trying to diversify by gender and age,[15,16] They were formed around issues on which everyone could agree and these were eventually combined into larger committees. However, the organisation noted that there were several issues with creating and sustaining committees. Firstly, Syrians and Lebanese had different ideas of 'community'. Particularly for Syrians, who moved because of jobs or because they could not pay rent, their idea of neighbourhood and community would be qualitatively different from their

Lebanese counterparts.[17] Further, the deteriorating socio-economic situation in Lebanon also led to tensions between different groups.[18] In fact 'communities' varied in different neighbourhoods- being fully formed in some places and non-existent in others. This lead to different outcomes in terms of committee empowerment and project success[19] especially as organisations found themselves not only attempting to create social cohesion through committee/community building, but training them in 'expert talk' and learning how to take grievances to the local authorities.[20] This latter issue of community creation and training was also affected by the short project cycles determined by the donors.

The TTC also concerned themselves with whom to include and exclude when thinking about neighbourhood makeup. The last draft notes that "The geographical scope of the neighbourhood will be **defined by** both spatial and social factors and agreed approximate size ... The factors will include **shared** needs, experience, identity, background, services, equal vulnerabilities, common interest in upgrading with **space** as the structuring point of entry, not country of origin."[21] Building on this, the committee agreed to have community representatives, focal points, key individuals be an integral part of the consultation process. It was also advised to have representatives from all relevant groups present in consultations. As the document developed, discussions over exit strategies and handing over projects became more salient. The concerns around community participation, maintenance, involvement of the municipality itself became key issues. Communities needed to take over the ownership of projects especially after NGOs had finished the physical upgrading work. This meant that not only would the community need to participate actively in the process, but that they needed to have to necessary knowledge and enthusiasm to do so. To address the first part of the problem, there would have to be capacity - building undertaken by intervening organisations to ensure that adequate training and development had taken place. To encourage the second part of the problem to be resolved, it was also necessary that municipalities be involved from the outset.

There are a number of issues that arise here with regards to the understanding and empowerment of communities. Firstly, although there was an acknowledgment of the highly varied nature of residents and concerns over elite capture of consultations, the drafts did not include discussions on power dynamics between different groups of people, or the questionable nature of consensus. This is even more complex within an environment such as Lebanon where the Syrian crisis has led to hostility between the Lebanese and Syrians and as noted above, increasing illegalisation, destitution, and invisibility of the Syrian population. In light of this, it can be asked to what extent, can Syrians, who are seen as guests in the country and are often in a vulnerable legal condition able to meaningfully participate in a consultation? How can they be considered part of the community if they are not there for significant periods of time? How do you draw together their priorities and those of the local communities which may be considerably different from each other?[22] Furthermore, drawing on the critical work on communicative planning, in situations where there are deepening differences between people, can the discussions be frictionless? Does placing faith in deliberative processes enable socially just outcomes or can it, as some scholars point out, bury possibilities of redistribution and enable injustice and continuing disparities between groups to take place (Fraser, 2000; Watson, 2006)?

In later interviews in 2018, it came to light that concerns around community involvement were apparently discussed in other meetings but not included in the guidelines because of the sensitivity the Lebanese government had toward the question of refugees. The government viewed attempts at advocating for refugees to be part of the community and participating equally as paths to have them naturalised. In a country where the presence of refugees has been thorny and seen to be destabilising the delicate political confessional balance, it is perhaps unsurprising that including them in participatory activities would be viewed with suspicion. All this points to the politically charged and highly unequal nature of participation and the ways in which humanitarian projects, much like development projects are capable to overlooking and silencing voices in the process of getting communities to participate (Mosse, 2004). Neighbourhood Approaches, though geographical and in theory comprehensive, may perhaps be seen as enabling a form of right to the city for refugees. However, it is important to reflect more critically on this perspective as it becomes evident that the involvement of displaced populations is always contingent and fleeting, driven by donor politics that are themselves fickle.

Equally surprising is the single mention of sectarianism in the evolving guidelines.[23] As noted earlier, the sectarian issue is significant in understanding both urbanisation and considering the Syrian crisis as neighbourhoods have either supported or rejected Syrians based on sectarian politics that spill over from Syria. In Tripoli for example, sectarian conflicts between adjoining neighbourhoods have complicated the work of NGOs doing neighbourhood level work.[24] It is only from May onwards that the TTC drafts contain language around conflict as well, and consideration for the conflict history of Lebanon. Suggestions on including this however extend to drawing on conflict maps that cover issues over the last two years, rather than thinking about conflict as a determining process in neighbourhood formation. Again, follow-up conversations with key informants revealed that although sectarianism was not mentioned in the drafts, it was part of the considerations. The lack of inclusion was also influenced by the involvement of a government representative as a co-lead of the TTC. However, not including these and other discussions in the text also meant not committing to specific guidance and allowing flexibility to adapt responses in the different local contexts .[25]

Finally, perhaps the most critical issues that complicate the development of urban responses for aid agencies are the geopolitical and geo-financial landscapes of humanitarianism. For example, the underlying concern was to ensure that upgrading and other projects would be appealing to donors, and manageable for organisations to implement within their limited mandates and budgetary constraints. This was being done against a backdrop of unpredictable funding, particularly in the shelter sector. UNHCR's budget fluctuated in different years and NGOs had to reach out to different donors for funding their assessments and projects. Donors ultimately determine the nature and length of funding and the limited funding cycles (sometimes only a year) affected the work of organisations working in vulnerable neighbourhoods and communities as they could not employ staff or sustain engagement in meaningful ways (Campbell, 2020). One interviewee from an INGO was particularly frustrated and noted that really the projects were donor driven, not needs driven - that donors would change priorities every year, from shelter, to WASH to development or something else.[26] In subsequent discussions they also noted that at the end of the day, they were a business, and the project cycle and donor priorities affected their long-term work and especially the engagement they hoped to have with the neighbourhoods and communities in which they were working.[27] Convincing donors then of shifting away from a refugee-focused, project cycle-based aid system to one that is inclusive and longer-term is difficult. Donors such as the European Union's European Civil Protection and Humanitarian Aid Operations (EU ECHO) were initially unwilling to support neighbourhood assessments as they saw it outside of the scope of humanitarian intervention[28] and many donors continue to refuse funding because projects are seen as being "not humanitarian enough" and "too development".[29] The Lebanese government was also reticent about coming on board with the neighbourhood plans, especially as it involved the participation of refugees as well. Further, different municipalities had different approaches to hosting refugees with some being more restrictive in terms of their rights than others (Barjas, 2016; El Helou, 2014; Sanyal 2018).

## Conclusions

The Neighbourhood Approach as a discrete policy failed to materialise due to a number of reasons, including sensitivities by the Lebanese government that this was perhaps a backdoor to permanently settling Syrians in the country.[30] This in itself is an interesting point to consider when unpacking how local policies are determined and how global policies fail to materialise. Donor politics are also key here- an insistence on focusing on project cycles and concrete outcome flies in the face of these area-based approaches that are intended to be medium-term with far more "soft" components.

What this brief glimpse into the policy-making process reveals is the deep entanglement of the local and the global in its formulation. Certainly, the Neighbourhood Approach is a key new global policy toolkit, but its emergence within the Lebanese context is not just about adapting global policy to a local context, but also about how the local context demanded this policy, and how in arriving, the policy folded territories from elsewhere and within. Tripoli, Beirut, Mount Lebanon shaped the policy through the experiments being done in them. This returns us then to Robinson's point that the provenance of policies is perhaps less interesting than thinking through what compels the production of certain polices and ultimately thinking about the impossible separation, the topological and the topographical relationship between different places in the making of policies.

Despite the effort that went into it, it was ultimately the local geographies and politics combined with the global aid landscape that determined the future of this neighbourhood policy. In this case, although the policy itself failed to materialise in its intended form, it has continued to be relevant within the context of Lebanon. Individual NGOs continue to use neighbourhood approaches in their work and some projects of upgrading houses, installing street lighting, improving safety have been completed in a number of neighbourhoods across Lebanon.[31] The Neighbourhood Approach as undertaken by some of these organizations have also become their own unique models and 'travelled' to other sites, such as Turkey and discussed in policy settings as an important template.[32] Aspects of it have also been continued by various UN agencies. For example, UN Habitat has created urban profiles for cities such as Tripoli and Tyre and "some early stages of inter-sector coordination at the city level was initiated, allowing for more focused coordination" (Bergby, 2019, p. 110). Profiling continues to be developed in collaboration with UNICEF and UN Habitat and they are being considered as templates for action by local communities and organisations.[33] This raises the possibility of a new conceptual understanding of neighbourhood, one that is driven by humanitarian logics and finance but continually remade on the ground. What future does this development of neighborhood profiles and policies hold for Lebanon and for elsewhere where such experiments may be occurring? How could these processes challenge earlier ideas of neighbourhoods?

Finally, the paper offers a challenge to the some of the work on policy mobilities such as that by Peck & Theodore(2010) that focuses primarily on urban entrepreneurialism and urban political economy and notes that Vancouver and Barcelona have purchase within the policy circuits because of their policy fixes. They note that "policy blogs are unlikely to be running hot, anytime soon, with talk of the Havana model, Kabulism, or even lessons from Detroit " (ibid, pg 171). I question this position. Cities become sites of learning for myriad reasons. Lebanese cities such as Tripoli and Beirut are rightly key sites for learning how to cope with and emerge from multiple and overlapping crises, of charting a path through global and local politics and priorities. They become models for the growing numbers of cities facing conflicts, disasters, and displacement. Perhaps a focus on the making and mobilisation of policies in these sites can serve as a way to provincialise conversations around policy mobilities as well.

## Notes

1. As noted by an anonymous committee member, The TTC was formed from the Shelter Working Group specifically to develop the guidelines for the Neighbourhood Approach.
2. The cluster approach was created in 2005 by the interagency standing committee (IASC) of the UN. "Clusters are groups of humanitarian organizations, both UN and non-UN, in each of the main sectors of humanitarian action, e.g., water, health and logistics. They are designated by the IASC and have clear responsibilities of coordination". See https://www.humanitarianresponse.info/en/coordination/Clusters
3. See https://www.uclg.org/sites/default/files/cities_in_crisis.pdf
4. Comments from TTC drafts.
5. Internal NGO report
6. Examples include CARE International's "One Neighborhood Approach", Solidarities International "El Hay" and ACTED's "Neighborhood Approach".
7. Interviews undertaken with RA in 2015.
8. Interviews undertaken with RA in 2015.
9. Comments by anonymous committee member
10. Notes from RA and e-mail correspondence
11. KI interview, June 2020
12. Notes from anonymous committee member
13. KI interview, June 2020
14. Notes from RA
15. The composition of committees could be very varied depending on how and by whom it was constructed.
16. KI interviews in May and June 2020. They also noted that men couldn't be on committees because of work and also because they didn't want to.
17. KI interview, May 2020
18. Interview with LNGO undertaken with RA, August 2015
19. Internal NGO report
20. KI interviews in May and June 2020

21. Emphasis mine
22. KI interview, May 2020
23. Notes from anonymous committee member
24. KI interview, June 2020
25. KI interview, 2018
26. Interview undertaken with RA, April 2015
27. Discussions with KI in 2017 and 2020
28. Notes from anonymous committee member
29. KI interview, June 2020
30. Interview with anonymous committee member, 2017 and 2020
31. See for example, publicly available project reports from organizations such as CARE Lebanon.
32. See for example, CARE Lebanon's "One Neighborhood Approach"
33. For a more detailed discussion regarding the methodological issues and adoption of elements of the Neighbourhood Approach, please see (Bergby, 2019).

## Acknowledgments

Part of the data was generated from the project "Policy Mobilities, Urban Politics and Assemblage: A View from Lebanon (2014-2017)," part of the London School of Economics and Political Science (LSE) Academic Collaboration with Arab Universities Programme funded by the Emirates Foundation and administered by the LSE Middle East Centre. Ms Jessy Nassar was the Research Assistant on the Project from March 2015- August 2016. I would like to thank the editors Rene Kreichauf and Birgit Glorius for inviting me to be part of this special issue and for their comments and support through the process. Additional thanks go to Susan Moore and three anonymous reviewers for constructive feedback on this paper. Finally, I am very grateful to Synne Bergby, Daniel Delati and the NGOs in Lebanon for generously sharing their experiences and knowledge, and to Jonathan Darling, Alan Mace, Eugene McCann and Jennifer Robinson for their generous and incisive feedback that helped strengthen this article.

## Disclosure statement

No potential conflict of interest was reported by the author.

## Funding

This work was supported by the Emirates Foundation.

## References

Abraham, Janaki. (2018). Exploring the contours of legitimacy in neighbourhoods in North Kerala, India. Urbanities-Journal of Urban Ethnography, *8*(1), 32–37. http://www.anthrojournal-urbanities.com/wp-content/uploads/2018/04/9-Abraham.pdf

Atallah, Sami. (2016, February). *Can municipalities take on the refugee crisis? Lebanese Centre for Policy Analysis* (blog). http://lcps-lebanon.org/featuredArticle.php?id=68

Barjas, Elham. (2016, September 30). *Restricting refugees: Measuring municipal power in Lebanon*. The Legal Agenda. http://legal-agenda.com/en/article.php?id=3174

Bergby, Synne. (2019). The role of planning in humanitarian response, looking at urban crisis response in lebanon. Edited by Libby Porter. *Planning Theory & Practice, Borders and Refuge: Citizenship, Mobility and Planning in a Volatile World*, *20*(1), 105–111. https://doi.org/10.1080/14649357.2018.1558566

Blokland, Talja. (2009). Celebrating local histories and defining neighbourhood communities: Place-making in a gentrified neighbourhood. *Urban Studies, 46*(8), 1593–1610. https://doi.org/10.1177/0042098009105499

Bou Akar, Hiba. (2018). *For the war yet to come: Planning beirut's frontiers* (Illustrated Edition ed.). Stanford University Press.

Boustani, Marwa, Carpi, Estella, Gebara, Hayat, & Mourad, Yara. (2016). *Responding to the Syrian Crisis in Lebanon: Collaboration between aid agencies and local governance structures.* International Institute for Environment and Development. JSTOR. https://www.jstor.org/stable/resrep02719

Brenner, Neil, & Theodore, Nik. (2002). Cities and the geographies of 'actually existing neoliberalism'. *Antipode, 34*(3), 349–379. https://doi.org/10.1111/1467-8330.00246

Campbell, Leah. (2018, April 4). *What's missing? Adding context to the urban response toolbox | ALNAP.* https://www.alnap.org/help-library/whats-missing-adding-context-to-the-urban-response-toolbox

Campbell, Leah. (2020). *One neighbourhood: CARE's humanitarian response in tripoli: ALNAP case study.* ODI/ALNAP.

Cowie, Paul, & Davoudi, Simin. (2015). Is small really beautiful? The legitimacy of neighbourhood planning. In Simin Davoudi & Ali Madanipour (Eds.), *Reconsidering Localism* (pp. 68–89). Routledge.

Darling, Jonathan. (2017). Forced migration and the city irregularity, informality, and the politics of presence. *Progress in Human Geography, 41*(2), 178–198. https://doi.org/10.1177/0309132516629004

Deeb, Lara, & Harb, Mona. (2013). *Leisurely Islam: Negotiating geography and morality in Shi'ite South Beirut.* Princeton University Press.

El Helou, Maya. (2014, December 22). *Refugees under curfew: The War of Lebanese municipalities against the poor.* Legal Agenda. http://legal-agenda.com/en/article.php?id=3052.

Fawaz, Mona. (2008). An unusual clique of city-makers: Social networks in the production of a neighborhood in beirut (1950–75). *International Journal of Urban and Regional Research, 32* (3), 565–585. https://doi.org/10.1111/j.1468-2427.2008.00812.x

Fawaz, Mona. (2009). Neoliberal Urbanity and the Right to the City: A View from Beirut's Periphery. *Development and Change, 40*(5), 827–852. https://doi.org/10.1111/j.1467-7660.2009.01585.x

Fawaz, Mona. (2016, May). Planning and the refugee crisis: Informality as a framework of analysis and reflection. *Planning Theory,* 16(1), 1473095216647722. https://doi.org/10.1177/1473095216647722

Fawaz, Mona, Saghiyeh, Nizar, & Nammour, Karim. (2014). *Housing, land and property issues in Lebanon: Implications of the Syrian refugee crisis.* UN Habitat and UNHCR. http://unhabitat.org/housing-land-and-property-issues-in-lebanon-implications-of-the-syrian-refugee-crisis-august-2014/

Fraser, Nancy. (2000). Rethinking Recognition. *New Left Review, 3*(May/June), 107–120. https://newleftreview.org/issues/ii3/articles/nancy-fraser-rethinking-recognition

Gabiam, Nell. (2016). Humanitarianism, development, and security in the 21st century: Lessons from the Syrian refugee crisis. *International Journal of Middle East Studies, 48*(2), 382–386. https://doi.org/10.1017/S0020743816000131

Government of Lebanon. (2017). *Lebanon crisis response plan (LCRP) 2017-2020 - full version.* Lebanon. https://data2.unhcr.org/en/documents/details/53061

Government of Lebanon, and United Nations. (2014). *Lebanon crisis response plan 2015-16.* Lebanon. http://www.un.org.lb/english/lcrp

Harb, Mona, & Atallah, Sami (eds). (2015). *Local governments and public goods: Assessing decentralization in the Arab World.* The Lebanese Center for Policy Studies. http://www.lcps-lebanon.org/publications/1446627069-1443012809-decentralisation_latin_web.pdf

Harb, Mona, Kassem, Ali, & Najdi, Watfa. (2018). Entrepreneurial refugees and the city: Brief encounters in beirut. *Journal of Refugee Studies.* 32(1), 23-41. https://doi.org/10.1093/jrs/fey003

Holman, Nancy, & Rydin, Yvonne. (2013). What can social capital tell us about planning under localism? *Local Government Studies*, *39*(1), 71–88. https://doi.org/10.1080/03003930.2012.675330

Kallus, Rachel, & Law-Yone, Hubert. (2000). What is a neighbourhood? The STRUCTURE AND FUNCTION OF AN IDEA. *Environment and Planning. B, Planning & Design*, *27*(6), 815–826. https://doi.org/10.1068/b2636

Khechen, Mona. (2018). The remaking of ras beirut. *City*, *22*(3), 375–395. https://doi.org/10.1080/13604813.2018.1484643

Kihato, Caroline Wanjiku, & Landau, Loren B. (2016, October). Stealth humanitarianism: Negotiating politics, precarity and performance management in protecting the urban displaced. *Journal of Refugee Studies*, 30(3), 407-425. few031. https://doi.org/10.1093/jrs/few031

Lord, Alex, & Tewdwr-Jones, Mark. (2014). Is planning 'under attack'? Chronicling the deregulation of urban and environmental planning in England. *European Planning Studies*, *22*(2), 345–361. https://doi.org/10.1080/09654313.2012.741574

Martin, Diana. (2015). From spaces of exception to 'campscapes': Palestinian refugee camps and informal settlements in beirut. *Political Geography*, *44*(January), 9–18. https://doi.org/10.1016/j.polgeo.2014.08.001

McCann, Eugene. (2011). Urban policy mobilities and global circuits of knowledge: Toward a research agenda. *Annals of the Association of American Geographers*, *101*(1), 107–130. https://doi.org/10.1080/00045608.2010.520219

Mosse, David. (2004). *Cultivating development: An ethnography of aid policy and practice* (4th ed.). Pluto Press.

NRC ICLA. (2014). *The Consequences of limited legal status for syrian refugees in lebanon, part 2: NRC Lebanon field ASSESSMent, North, Bekaa and South*. Norwegian Refugee Council. http://www.nrc.no/?aid=9147816

Nucho, Joanne Randa. (2016). *Everyday Sectarianism in Urban Lebanon: Infrastructures, public services, and POWer*. Princeton University Press.

Parker, Elizabeth, & Maynard, Victoria. (2015). *Humanitarian response to urban crises a review of area-based approaches*. IIED. https://pubs.iied.org/pdfs/10742IIED.pdf

Parker, Gavin, & Street, Emma. (2015). Planning at the neighbourhood scale: Localism, dialogic politics, and the modulation of community action. *Environment and Planning. C, Government & Policy*, *33*(4), 794–810. https://doi.org/10.1068/c1363

Peck, Jamie, & Theodore, Nik. (2010). Mobilizing policy: Models, methods, and mutations. *Geoforum, Themed Issue: Mobilizing Policy*, *41*(2), 169–174. https://doi.org/10.1016/j.geoforum.2010.01.002

Prince, Russell. (2017). Local or global policy? Thinking about policy mobility with assemblage and topology. *Area*, *49*(3), 335–341. https://doi.org/10.1111/area.12319

Robinson, Jennifer. (2013). 'Arriving at' urban policies/the urban: Traces of elsewhere in making city futures. In Ola Söderström, Shalini Randeria, Didier Ruedin, Gianni D'Amato, & Francesco Panese (Eds.), *Critical mobilities* (pp. 23–50). EPFL Press.

Robinson, Jennifer. (2015). 'Arriving at' urban policies: The topological spaces of urban policy mobility. *International Journal of Urban and Regional Research*, *39*(4), 831–834. https://doi.org/10.1111/1468-2427.12255

Robinson, Jennifer. (2018). Policy mobilities as comparison: urbanization processes, repeated instances, topologies. *Revista De Administração Pública*, *52*(2), 221–243. https://doi.org/10.1590/0034-761220180126

Sanderson, David. (2017). Implementing area-based approaches (ABAs) in urban post-disaster contexts - David Sanderson, 2017. *Environment and Urbanization*, *29*(2), 349–364. https://doi.org/10.1177/0956247817717422

Sanderson, David. (2019). Humanitarian response in urban contexts. *Good Practice Review*, *12*, 244. https://odihpn.org/wp-content/uploads/2019/03/GPR-12-2019-001-244-web4_FINAL.pdf

Sanderson, David. (2020). Urban response: Three principles for good practice. *Forced Migration Review*, *63*(February), 8–10.

Sanyal, Romola. (2017). A no-camp policy: Interrogating informal settlements in Lebanon. *Geoforum, 84*(August), 117–125. https://doi.org/10.1016/j.geoforum.2017.06.011

Sanyal, Romola. (2018). Managing through Ad Hoc Measures: Syrian refugees and the politics of waiting in Lebanon. *Political Geography, 66*(September), 67–75. https://doi.org/10.1016/j.polgeo.2018.08.015

Sanyal, Romola. (2014). Urbanizing refuge: Interrogating spaces of displacement. *International Journal of Urban and Regional Research, 38*(2), 558–572. https://doi.org/10.1111/1468-2427.12020

Sawalha, Aseel. (2010). *Reconstructing beirut: Memory and space in a postwar arab city*. University of Texas Press.

Schell, James, Hilmi, Mohammed, & Hirano, Seiko. (2020, February). Area-based approaches: An alternative in contexts of urban displacement. *Forced Migration Review*, 63 (February), 16-19. http://www.fmreview.org/cities

Seethaler-Wari, Shahd. (2018). Urban planning for the integration of refugees: The importance of local factors. *Urban Planning, 3*(4), 141–155. https://doi.org/10.17645/up.v3i4.1696

Seidman, Steven. (2012). The politics of cosmopolitan beirut: From the stranger to the other. *Theory, Culture & Society, 29*(2), 3–36. https://doi.org/10.1177/0263276411410446

Stein, Rachel E. (2014). Neighborhood scale and collective efficacy: Does size matter? *Sociology Compass, 8*(2), 119–128. https://doi.org/10.1111/soc4.12127

Watson, Vanessa. (2006). Deep difference: Diversity, planning and ethics. *Planning Theory, 5*(1), 31–50. https://doi.org/10.1177/1473095206061020

World Bank. (2016). *The welfare of syrian refugees: Evidence from Jordan and Lebanon*. https://doi.org/10.1596/978-1-4648-0770-1

# Refugees and the transforming landscapes of small cities in the US

Pablo Shiladitya Bose

**ABSTRACT**
Over the past thirty years, immigrants in the US have been settling outside of gateway cities, going instead to a number of other new locations across the country. While much of the extant literature has focused on economic and family-based migration, in this paper I look at a different subgroup, officially resettled refugees. I explore the rationales that lie behind such a trend by examining placement data and a number of interviews with key figures in the US resettlement system at multiple levels. In particular, I examine the idea of refugees as an instrument of urban revitalization for smaller cities as well as the opportunities and challenges that their arrival has presented for these sites. I contextualize the arrival of refugees in smaller cities in the US within the current moment of rising racism, xenophobia and anti-immigrant sentiment across the country, as well as a resettlement program undergoing radical changes..

## 1. Introduction – refugees and urban change in small cities

Since the US Refugee Admissions Program (USRAP) was officially inaugurated in 1980, several million refugees have been resettled in the US, from a diverse range of countries and regions. Most of them were placed by the federal government and its non-governmental resettlement agency partners in the major metropolitan areas that immigrants have traditionally favored in the US – gateway cities like Chicago, New York and Los Angeles (Portes & Rumbaut, 2014; Singer & Wilson, 2011). Yet over the past three decades, much as broader immigrant settlement patterns have seen a shift in newcomers arriving in newer locations – in the Midwest, the South, rural areas and smaller towns and cities (Massey, 2008; Singer et al., 2008) – we can see similar patterns with refugee placements, with today's newcomers settling in increasing numbers in diverse locations: secondary and smaller cities, suburban areas, rustbelt regions, Midwestern and Southern states, and in rural areas all over the country (RPC, 2018). Even in the Northeastern US – a traditional destination for many earlier (primarily European) immigrant groups – the influx of refugees from different parts of Africa, Asia and the Middle East has profoundly shifted the demographics and social and political life of many communities (Coughlin et al., 2016).

Such trends have themselves undergone another significant shift since 2017, when the Trump administration took office. As part of its efforts to radically reshape the nature and structure of immigration in the US – both legal and irregular – the current administration has sought to curtail and change refugee resettlement. This has included attempting to suspend the program entirely (as part of several proposed "travel bans"), drastically cutting the numbers of approved refugees to be resettled, lengthening and making more difficult the approvals process overseas, and cutting funding to both government programs serving refugees and to the partner agencies who help to resettle newcomers (Aldana, 2018; Annett, 2017; Brandt, 2017; Karas & Campbell, 2020; Morello, 2018).

This is of course not the first time that refugee resettlement in the US has come under assault; it has seen a number of controversies in the years since its formalization as a primarily federal program in 1980. While some groups have been subject to considerable backlash regarding their admittance, others have been given preferential treatment (Portes & Rumbaut, 2014; Timberlake & Oahn Cook, 1984). By the beginning of the twenty-first century, however, the US had emerged as one of the most significant third country resettlement destinations in the industrialized world – in absolute numbers, if not per capita (UNHCR, 2019). But since 2015, refugee resettlement as a concept, a process and a practice has come under considerable criticism and constraint not only in the US but worldwide. While refugees have always been a fairly small part of the overall immigration landscape globally, they have been thrust onto center stage for a variety of reasons in recent years. Conflicts in the Middle East and North Africa have combined with environmental disasters, droughts, famines and economic instability to send millions fleeing across borders toward wealthier countries in Europe and North America. At the same time anxieties in receiving countries of the Global North regarding long-term demographic change and a more immediate influx of migrants, have coupled with growing Islamophobia and paranoia sparked by terrorist attacks carried out by Islamic extremists across the globe – often blamed on refugees and asylum seekers but carried out for the most part by native-born or immigrant (but not refugee) individuals and groups – to create a significant backlash against refugees worldwide.

In the US, multiple state governors and a smaller number of cities announced in November 2015 (following attacks in Paris) that they would no longer support refugee resettlement in their areas. The issue became a central feature of the US presidential campaign during 2015 and 2016, with then-candidate Donald Trump taking a leading role in advocating against refugee resettlement, especially from Muslim-majority countries. In 2016 states like Texas and Maine officially withdrew from the USRAP, though refugees were still resettled in those states, especially Texas which has the largest share of refugee arrivals in the US (Kennedy, 2016). And the first three years of the Trump administration has ushered in an era of uncertainty and anxiety for the USRAP, manifested through multiple cuts and changes to the refugee resettlement program.

Despite these changes and the existential crisis facing the USRAP, many millions of refugees have been resettled in the US over the past forty years. In this paper I focus on their arrival in non-traditional immigrant destinations, on the rationales for being sent to such locations, local attitudes toward their arrival and on the role of refugees in ongoing processes of urban change. I argue that what we see in this dynamic is a clash between competing sets of visions of what the future holds for such urban spaces. On the one hand is the idea that refugee resettlement is a key way to transform certain kinds of smaller cities in the US – rustbelt or aging cities, for example. Resettling refugees from this perspective can make such cities more diverse, younger in population, provide new workers, homeowners, and businesses or simply add residents. Contrasted to this positive view is one in which refugee arrivals manifest a set of anxieties about social, cultural, economic, and demographic changes. Some of these fears may be based in ignorance, xenophobia and intolerance, yet others reveal deeper discomfort about the ways that cities – especially those that are smaller, secondary or located in rural areas – chart their ways forward during a period of transition. In this paper I look at both sides of this dynamic and ask what if any role refugees play in the changing nature of their new homes.

The arguments in this paper are drawn from a larger project looking at the transformation of refugee resettlement in the US and globally over the past decade, with particular attention to non-traditional immigrant destinations. In this paper I discuss the methods used to collect the empirical data on which the arguments are based, some of the relevant literature on refugees and resettlement in urban sites (with a primary focus on the US experience), the context of refugee resettlement in the US, the rationale for placing refugees in newer and smaller locations, and the reactions to such settlement. I argue that the case for placing refugees in smaller cities and non-traditional locations in the US is based on the idea that they add diversity to relatively homogenous populations and that they spur revitalization – or even represent a last chance for survival – for urban spaces in decline. Yet I also ask what the potential for such hopeful narratives might be in the context of anti-immigrant sentiment and government policy and the evisceration of the US refugee resettlement program in the current moment.

## 2. Refugees and immigrants in new urban destinations

While the term "refugee" is often used as a catch-all to describe an entire range of those forced to flee their homes – by conflict, persecution, economic deprivation, environmental change, or by other factors – within international law and practice, the scope of who can claim the protection offered by the label is much narrower (Dauvergne, 2008). It is mostly those who have been able to prove – over much time and with much difficulty to a wide range of authorities and institutions – a credible fear of harm through various forms of persecution or conflict in their home countries, have most often been housed in a refugee camp in another country and have been subsequently granted resettlement in a third country who can claim official refugee status through international law and its attendant mechanisms (Loescher et al., 2008).

There are many others who fall under the category of forced migrant – internally displaced within a country or those displaced by development projects and environmental causes, for example, – yet official refugees are the most closely scrutinized and thus most carefully vetted of such groups (Haines & Rosenblum, 2010). In the US, they have had to have their status approved while still abroad, as opposed to asylum seekers, whose reasons for forced migration may be similar but who claim sanctuary once arrived in their hoped-for host country. Within the US there are several other categories through which forced migrants might arrive – asylum seekers, special immigrant visa holders,[1] or those with temporary protected status[2] or unaccompanied minors.[3] But for the purposes of this paper, I will be focusing on refugees arriving in the US through the UNHCR-directed global refugee processing system.

Much of the work on refugees and resettlement more recently has focused on urban camps (Beier & Fritzsche, 2017; Kallio et al., 2019; UNHCR, 2009), while another literature looks at refugees in gateway cities (Danso, 2002; Gow, 2005; Sanyal, 2012; Tang, 2015). The literature on resettlement in non-traditional immigrant destinations is more limited but offers several helpful insights for the cases considered in this paper – in particular on the role that such locations play in the acculturation process and the impacts that refugees might have on smaller resettlement sites.

When thinking about the experiences of refugees arriving in new locations, Miyares (1998) research on Hmong resettlement in Merced and Fresno, CA, for example, suggests that Christianity and education were among the chief determinants of positive outcomes for the new immigrants. Jones's (2008) Immigrants Outside Megapolis in his examination of resettlement in small communities in Wisconsin, Iowa, Louisiana and New York argues that successful experiences seem predicated on the pooling together of inter- and intra-community resources. Kraly and Vogelaar 's (2011) work on Utica, NY is an example of another common theme in resettlement in smaller communities, as the authors detail the impacts of the significant influx of unfamiliar newcomers (Africans and Asians); in particular the effects on multiple neighborhoods, including the growth of new businesses, new housing starts, and establishment of vibrant community networks.

Indeed, the positive experiences in some of these new(er) immigrant destinations often highlights the challenges of resettlement in larger, putatively better prepared sites. For example, Newberry and Darden (2011) argue that the Somali experience of being placed in Atlanta – often in the midst of high-crime neighborhoods – has arguably been worse than for those placed in small towns in the Northeast. At the same time, Singer and Wilson (2011) suggest that the emergence of larger rustbelt cities as new destinations and the idea of refugees and immigrants as a supposed catalytic force for economic revitalization has been a mixed proposition, with newcomers often restricted to neighborhoods in which there are less opportunities for upward social mobility. For example, as Hardwick (2011) points out refugees from both Eastern Europe and Africa placed in Portland, OR, found themselves spatially segregated in housing patterns similar to those experienced by immigrant groups of the past.

This is not to suggest however that the scale of smaller sites will automatically guarantee a smooth transition for refugees. For example, Pho et al. (2007) collection of essays on Southeast Asians in Lowell provides some interesting insights with a longer perspective. Looking back at nearly forty years of resettlement in a postindustrial New England town the authors examine the very different experiences of Cambodians, Laotians, and Vietnamese refugees based on their ethnic and class backgrounds, religious networks, and experiences of trauma. The authors suggest that for a small town the level of unfamiliarity with the incoming population may lead to considerable conflict over such issues as support for languages other than English in the school system, trauma counseling that is culturally appropriate, or political representation. Such a situation is not unique to Lowell. There are many instances of misunderstandings and frustrations regarding newcomers boiling over – in one of the best-known cases in Lewiston, ME, the influx of Somali refugees led to its mayor famously urging them in 2002 to stop coming any further and spare overburdened city services. As Rabrenovic (2007) details, the mayor's actions (and the subsequent attempts to exploit the situation on the part of white supremacist groups) was at considerable odds with what had actually been an effective collaboration between the Somali community, resettlement agencies, and local officials to manage the resettlements and led to feelings of betrayal on the part of many refugees. Despite the high-profile nature of this debacle, the actual placement of Somalis in Lewiston has proved to be successful and a benefit to all residents, not just refugees. In Rabrenovic's view

> In Lewiston, immigrants contributed to the city's economic and social transformation and made it into a more vibrant and worldly place. Also, although the initial impetus to improve delivery of municipal services was based on meeting the needs of the immigrant population, other residents, such as elderly and residents with special needs, benefited as well. Thus, inter-dependence between newcomers and established residents developed (Rabrenovic, 2007, p. 358)

The issue of refugee acculturation into small and mid-sized towns is therefore about more than just the capacity to absorb newcomers. It is equally about the types of social and spatial formations and identities that refugees must learn about and adjust to. Of especial concern are the ways in which refugees adjust to life in often predominantly white small towns and cities – especially those newcomers who are marked as racially, religiously, or ethnically distinct even in places that have welcomed immigrants in the past. Other studies have shown that refugees, like other immigrants, are often subject to existing patterns and practices of spatial inequality such as segregation (Feng, 2006), marginalization (Eastmond, 2011), and social exclusion (Haines & Rosenblum, 2010; McBrien, 2011). A lack of access to employment and healthcare are a source of concern for many refugees in their new homes (Mitschke et al., 2011), as is participation in education (Neidell & Waldfogel, 2009). More broadly, Abdelkerim and Grace (2012) suggest that poor resettlement experiences lead to a lack of self-sufficiency and political agency amongst refugees, while Uteng argues that "constrained mobility [is] a constitutive factor of social exclusion" (Uteng, 2009, p. 1057) for refugee women in Scandinavia. Such studies would indicate that refugee lives – rather than following the much mythologized immigrant narrative of upward social mobility (Roediger, 2005) – are often challenged by existing structural inequities (Carter & Vitiello, 2011). This is especially borne out by statistics on housing – Carter and Osborne (2009) find, for example, that three years after arrival 85% of refugees settling in Winnipeg remain in the rental sector while work by Hiebert and Mendez (2008) uncovered almost negligible rates of ownership amongst refugee households in Vancouver four years after their arrival. In the newer destinations where refugees increasingly find themselves, such problems can become compounded:

> Although many Convention refugees have positive resettlement experiences in smaller cities and towns, others suffer from the absence of a critical mass of immigrants with similar cultural backgrounds and from limited outreach and miscomprehension by the local receiving communities. These newcomers may resort to frequent and costly trips to larger urban centres in order to access appropriate services and maintain their networks. (Rose & Ray, 2011, p. 150)

Sherrell et al. (2005) in their examination of the resettlement experiences of Kosovars in smaller sites in British Columbia found similar mixed results. On the one hand newer locations often provided greater flexibility in such things as expanded definitions of family that allowed more within extended networks to cluster in the same place. On the other hand, this led to a lack of interaction with the broader population, an inability to access language training and employment, and a lack of recognition of existing credentials. The authors suggest that while resettlement in new locations was a qualified success in that the refugees did not leave their initial placement sites, it was unclear whether small sites facilitated faster integration. The literature on refugee resettlement in newer destinations outside of the traditional gateway cities thus suggests a mixed record of successful outcomes and ongoing challenges, and the necessity for a nuanced approache as I undertook my own inquiry into resettlement experiences in the US.

## 3. Methods and sources

In this paper I focus on two sets of empirical evidence to look at refugees and resettlement in the US. The first is an analysis of a dataset of resettlement locations in the US between 2012–2016 (the second term of the Obama administration, when approvals for refugee resettlement had reached a "steady state" of 75,000 individuals per year) to identify the pace and scale of refugee arrivals in non-traditional metropolitan urban areas (primarily smaller cities and towns). This dataset, known as the Worldwide Refugee Admissions Processing System (WRAPS) was developed by the Refugee Processing Center of the US State Department. It tracks the movement of all refugees from around the world by country and region of origin and to the sites of initial settlement in the US, using the headquarters of the refugee resettlement agencies as a proxy for resettlement location.

I have used the WRAPS data to identify and discuss the urban sites for refugee resettlement during the study period. While WRAPS data gives a very good snapshot of resettlement during initial placements, it does not track secondary migration to other locations. It is, however, a useful tool to understand where and how the US government and its partners choose to send refugee arrivals; there is a strong incentive for refugees to remain in their initial resettlement location for at least the first year since their package of benefits is tied to that place, rather than to the individual or household and is not portable.

The second source I have used to examine the rationale for sending refugees to their particular host locations and to the reception they have encountered there is a set of interviews with key stakeholders within the assemblage of the refugee resettlement regime in the US, known as the US Refugee Admissions Program (USRAP). This includes over 250 interviews with officials within each of the federal agencies, state governments, resettlement organizations, ethnic associations, and municipal governments that comprise the USRAP. The structure of the USRAP is fragmented and often disjointed, a legacy of an earlier era in which resettlement was primarily led by faith-based, secular and ethnic or identity-based organizations advocating for the rights of refugees and asylum seekers and often leading integration efforts as private sponsors (Barneche & Joe, 2014). There are currently seven resettlement agencies (sometimes referred to as VOLAGs or voluntary agencies) with which the US federal government contracts, including Catholic, Lutheran, Episcopalian, Protestant, and Jewish umbrella groups, as well as secular and ethnically-based associations. Their fate is currently uncertain, as shrinking numbers of new arrivals and an administration hostile to immigration in general and to refugees in particular have made their future look bleak.

Over a six-year period, from 2012–2018, I conducted interviews with representatives from every state, each of the main actors at the federal level and during site visits at multiple sites across the US – cities of approximately 50,000–100,000 inhabitants in the metropolitan area within the broad sub-regions of the Northeast, Northwest, Midwest, South, and Southeast in the country. Each interview lasted approximately one hour and was semi-structured in nature. Interviews were transcribed, coded and analyzed using qualitative software in order to look for common patterns and themes in the perceptions of refugee resettlement policy, practice and experience within the USRAP during this time of transition and change.

One of the key challenges for my research in this time of high-profile controversies regarding refugee resettlement has been to maintain confidentiality and anonymity of respondents. Especially in smaller communities, the actual daily work of key informants may become constrained, complicated and even dangerous if their candid views were to be widely disseminated. For these reasons, respondents are not directly identified but are rather named in this paper using a broad job description (e.g., State Coordinator, Municipal Official, Federal Agency) and the general region in which they are located (e.g., Southwest, Northeast). Similarly, identifying details are masked within direct quotes in order to maintain anonymity. I have also not identified specific towns and cities involved in examples drawn from this research as masking their identity in some cases would be impossible. Such practices have a long history in the migration research (Dahinden, 2016; Jacobsen & Landau, 2003; Mackenzie et al., 2007).

## 4. Resettlement in the US

Before examining the case of resettlement in non-traditional destinations in the US, it is helpful to set out the context for and a brief history of refugees arriving in the country. What has been the more recent experience in the US? The refugee resettlement system that has come under such strong partisan attack is one that has evolved in the US since 1980 as a private-public-partnership, in which the US government approves the screening and selection of refugees overseas and a group of non-governmental organizations take charge of selecting placement locations and organizing the resettlement and integration process (Yan, 2006).

These agencies administer resettlement services in a number of different ways – in some cases they operate field offices as satellites while in others they sub-contract with local immigration organizations or social service providers. Cities, states and counties are thus also a part of the complex system that governs resettlement. Some states, for example, have extensive offices dedicated to refugee services and an official State Refugee Coordinator, while in others resettlement agency staff fulfill this statewide coordinating role. The deciding factors on how formalized refugee services may be in a given city, state, or county are not necessarily determined by the size of the population or even by the numbers of refugees being resettled in that location.

For example, California, the second largest resettlement state by the number of refugees approved for placement 2012–2016, has a dedicated Refugee Programs Bureau to coordinate its activities. On the other hand, Texas – the largest resettlement state during the same period – pulled out of the USRAP as a state partner in 2016 (Kennedy, 2016). Refugees still arrived in Texas (though as of 2020 they will no longer do so), but were supported primarily by nonprofit partners and municipal governments rather than the state government. Some of the traditional gateway cities such as New York, Chicago and Los Angeles may receive a relatively small number of officially resettled refugees and yet may have the infrastructure and experience to assist immigrants more broadly whereas for cities that are either smaller in terms of population or have less of a history receiving newcomers there is often a lack of dedicated staff or resources to such functions.

It is the decentralized nature of US resettlement that leads to such inconsistencies in the experiences of refugees. Altogether, this assemblage of federal, state and local authorities works to approve refugee admissions and (with the help of the resettlement agencies) helps to assign them to new homes. The USRAP also provides up to eight months of direct financial assistance to refugees and access to a range of social services (including employment aid, language training, medical care and food stamps) beyond the initial resettlement period. While the role played by government authorities is a key part of this process, some scholars have suggested that the parastatal functions taken on by the resettlement agencies have made them the formative institutions in the adaptation and integration experience for refugees (Clevenger et al., 2014; Nawyn, 2010). It is they who find refugees a place to live, a job, food and clothing, counseling, connections within their new community, language classes, and transportation to destinations. It is through them that the USRAP's priorities in integration – primarily focused on economic self-sufficiency – are often expressed. It is important to thus keep in mind the significant actors when considering the context of refugee arrival in new urban destinations – and it is why in my attempt to understand settlement patterns I have interviewed a range of key figures including state refugee coordinators, state health coordinators, federal officials, resettlement agency staff, municipal authorities, and community organizations and leaders.

As I state at the outset of this paper, the pattern of refugees being placed outside of the major metropolitan areas is not new and has been apparent since at least the early 2000 s (RPC, 2018). This trend echoes a similar one for immigrants more widely in the US – as numerous studies have shown that immigrants are settling outside of the gateway cities such as New York, Chicago and San Francisco that have long been associated with their arrival (Massey, 2008; Nelson et al., 2015; Singer et al., 2008). Secondary cities, small towns, and rural areas are increasingly host to newer groups of immigrants. Yet while labor migrants, those seeking family reunification, or those immigrating for educational opportunities have at least in theory some level of agency in choosing where it is that they might settle, refugees do not have the same choices. Their initial placement is instead chosen for them through the USRAP process (Mott, 2010). During my study period of 2012–2016, the trend of refugees being settled in smaller cities across the US was clearly apparent. While several traditional large metropolitan gateway cities like New York, Chicago and Los Angeles did participate in the USRAP, they did so at a much lower rate per capita than did many smaller communities. In Table 1 we can see, for example, while 10 of the 230 resettlement sites across the US were cities with populations of 1 million or more inhabitants, a far larger number of resettlement sites were in much smaller cities:

**Table 1.** Refugee resettlement sites in the US 2012–2016 by city population.

| Resettlement Site Population | Number of participating cities | Resettlement Site Population | Number of participating cities |
|---|---|---|---|
| 2 million + | 5 | 80 k-89 k | 5 |
| 1–1.9 million | 5 | 70 k-79 k | 6 |
| 800 k-999 k | 6 | 60 k-69 k | 17 |
| 600 k-799 k | 11 | 50 k-59 k | 13 |
| 500 k-599 k | 7 | 40 k-49 k | 17 |
| 400 k-499 k | 7 | 30 k-39 k | 9 |
| 300 k-399 k | 16 | 20 k-29 k | 9 |
| 200 k-299 k | 24 | 10 k-19 k | 9 |
| 100 k-199 k | 53 | Less than 10 k | 4 |
| 90 k-99 k | 7 | | |

Source: ACS (2013) and RPC (2018)

**Table 2.** Approved resettlements by site size and population 2012–2016.

| Resettlement Site Population Range | Number of Participating Cities | Total Approved Refugees | % of Overall Approvals |
|---|---|---|---|
| 1–2 million + | 10 | 72,964 | 18.45 |
| 400 k – 999 k | 31 | 116,925 | 29.57 |
| 100 k – 399 k | 93 | 138,030 | 34.91 |
| 70 k – 99 k | 18 | 15,608 | 3.98 |
| 40 k – 69 k | 47 | 31,322 | 7.92 |
| >10 k – 39 k | 31 | 20,392 | 5.16 |
| | | 395,391 | |

Source: ACS (2013) and RPC (2018).

There are several clusters discernible in the resettlement numbers during this period (Table 2).

What we can see here is that while just over 18% of refugees were approved for resettlement in the largest metropolitan areas in the US, nearly the same number were approved for placement in cities ranging from ten thousand to just under one hundred thousand inhabitants. In the latter locations, the impact of newcomers would be disproportionately greater. An important caveat to remember when looking at these groupings is that there are two distinct kinds of newer locations visible in these groupings. There are some smaller locations – especially in the Northeast, where metropolitan clustering is often much denser than in other parts of the country – that fall within the orbit of a much larger city or are part of a larger metropolitan statistical area.

A very different kind of smaller city participates in the USRAP as well, however – these are smaller cities that are not particularly close to large metropolitan locations. In Massachusetts and Pennsylvania, for example, while significant resettlement occurs in the Boston and Philadelphia areas, many refugees are placed in smaller cities at some distance such as Lowell and Springfield or Harrisburg and Lancaster. Similarly, in New York state, far more refugees are placed in Utica, Syracuse and Buffalo than in New York City. We can see similar patterns in most other regions in the US, with resettlement in large cities, in smaller cities in their orbit, and in smaller cities well outside of their metropolitan area.

## 5. Rationales and reactions

With this context in mind, I turn to the main focus of this paper. What is the intention behind placing refugees in smaller cities and towns within the US? As the literature previously reviewed suggests, the rationale is often that refugees will provide new opportunities for growth for such communities – that the new arrivals will open new businesses, purchase homes and improve neighborhoods, and strengthen the local labor pool as well as add to the municipal tax base. Such justifications were repeatedly echoed throughout my interviews with a wide range of key informants involved with refugee resettlement in the US. Out of some 250 respondents from federal, state and local resettlement agencies, several themes became apparent. I have chosen to focus on three that are most directly relevant to the topic of this special issue – the relationship between refugees and the city. What is it that refugees are offered to by a particular kind of city – in this case the smaller, non-traditional destination – and what, conversely, can such spaces offer to refugees? The three themes I focus on here are overlapping: 1) diversity, 2) revitalization, and 3) survival. I have broken down in the following table how many respondents in each of five categories indicated that these were among their top five reasons that refugees were being placed in smaller cities during my study period (the fourth and fifth most common themes cited by interviewees – but by much smaller margins – were housing availability and economic opportunity) (Table 3):

Housing and jobs both appeared in many interviewee responses as will be apparent momentarily – but as a part of the context or outcomes of the placement, rather than as a driver of it. In other words, respondents did not rate the rationale for putting refugees into a new destination as the availability of housing or the draw of a particular job – a fact explained in part by the fact that the job market or availability of affordable housing looks so different in different parts of the US. The cost of rental and housing markets in the Northeast or in coastal states was far higher than in the Midwest, while manufacturing and agricultural jobs were far more common in the latter than in the former. And yet despite regional variances in all other themes, there was remarkable unanimity across multiple scales and manifestations of the refugee regime in the US when it came to the other three main themes that echoed across my interviews with key informants. In the following sections, I explore these in greater detail

**Table 3.** Most common themes by interviewee.

| | State Refugee Coordinator | State Health Coordinator | Federal Agency | Resettlement Agency | Municipal Government | Community Organization | Total |
|---|---|---|---|---|---|---|---|
| Total # of 250 | interviewees | 50 | 50 | 15 | 30 | 50 | 40 |
| *Theme* | | | | | | | |
| Revitalization | 50 | 50 | 14 | 29 | 49 | 38 | 230 |
| Diversity | 50 | 50 | 13 | 29 | 39 | 29 | 212 |
| Survival | 43 | 42 | 14 | 24 | 48 | 32 | 202 |
| Housing | 36 | 32 | 10 | 21 | 26 | 30 | 155 |
| Economic | 32 | 30 | 8 | 26 | 30 | 28 | 154 |

### *5.1. Refugees as bringers of diversity to smaller communities*

Well, I think the impact in the smaller communities is that it also increases the community's idea of what's beyond themselves.

- State Refugee Coordinator, West

You know, for the community that's welcoming them, I think just the whole diversity that they bring to that situation. In addition to that, there's also that sense of just that newness they bring, because your community are either going to have a sense of wanting to help the refugees, or not. And certainly, for the ones that want to help, there's also that sense of contributing to that community as a result of helping out that population.

- State Refugee Health Coordinator, Midwest

We have other immigrants as well, but we are not a gateway city, we don't have – we're not a destination for many immigrants. We have some enterprises that do attract foreign workers but it's small numbers. We have a history of immigration, but it's small numbers. I think the refugee program, the refugee resettlement, the fact that we are resettling refugees adds significant cultural diversity and a more global view of the world. And it helps to revitalize the neighborhoods that they go into.

- State Refugee Coordinator, Midwest

Such statements raise a number of realities of many of the smaller cities and towns outside of larger metropolitan areas in the US. For many, their populations are relatively homogenous and have little in the way of histories of immigration – certainly for nonwhite immigrant groups. Refugees are thus an important way of changing the demographics of these cities. Why might such change be necessary? For a range of smaller cities and states across the US, some of the key issues that face them include aging populations, youth outmigration, the loss of local jobs and workers (and the foundations of a local tax base connected to them), and more broadly a shifting set of relationships to the global economy. Unlike larger metropolitan areas with longer histories of immigration that might additionally be more demographically diverse, many smaller communities may seem less attached to the outside world; in the US as with much of the rest of the world, this can feel like the separation of two solitudes, whether of a rural-urban or a political divide. For such reasons it is thus not surprising that a significant number of interviewees described the addition of refugees as a way of allowing smaller communities to widen their sense of self. This could take multiple forms – exposure to new languages, foods, religions and traditions for the majority population, the ability of children to encounter more difference in schools and within youth activities, and the possibility of connecting the experience of those living in smaller cities to the "wider world" as several respondents put it.

It is clear that those involved with the USRAP see this as a positive change; yet it is also clear that there are others who do not. One does not need to look as far back as the mid-2000 s in Maine or as high up as the Trump Administration to see the pushback at the municipal or state levels against resettlement. Instead, refugee newcomers have begun to engender much greater amounts of resistance. In Lewiston, for example, while Rabrenovic (and others) speak positively of the impact of refugee arrivals, a strong core of opposition remains (Misra, 2019). Changing demographics cannot alone address deeply embedded structures of racism so one must ask what if any resources are provided for those refugees placed within these new homes in order to navigate them.

Demographic transformations are certainly not seamless, especially when the groups growing in number are additionally marked by religious, racial and ethnic difference. As Chavez (2013) points out in the case of the increase in Latino labor migrant populations across the US in recent decades, this can lead to a backlash that depicts newcomers as a literal "threat" to jobs, safety, social cohesion, and cultural values. While proponents of refugee resettlement in smaller communities may see increased diversity as a boon, the debates that have raged nationwide regarding refugees since 2015 in particular have tended to focus on anxieties regarding difference – things like Sharia law, Muslim cultural practices and traditions, and unfamiliar languages. Refugees can be seen in this light as the embodiment of such fears.

Indeed, one of the major challenges facing the USRAP in the current moment is the sudden switch from a planned expansion of programs in 2015, when the federal government and its partners looked to open many new sites in smaller locations, to the contraction of the overall program and the closing of many of these same new sites by 2018 onwards under the Trump Administration. In some of these smaller locations, the possibilities of refugee resettlement have been overwhelmed by fears of demographic and cultural change, often expressed as security concerns regarding terrorism – if not outright ignorance and Islamophobia – and often local concerns have aligned with broader anti-immigrant politics in the US to sabotage resettlement plans (Keck, 2016).

## *5.2. Refugees as drivers of urban renewal*

If increasing demographic diversity alone is not enough of a reason to embrace refugee resettlement, key respondents' second theme seems a more generally straightforward and less cultural position to adopt – that refugees represent the opportunity to revitalize the cities and neighborhoods they settle into. The sentiments expressed by the previous quotes and the general sentiments of respondents are widespread – that refugees are drivers of urban renewal. They are responsible for improving specific areas in multiple ways. These include starting new businesses or providing more and newer consumers for existing ones. Refugees have also acted as a stable source of rental income for landlords in many smaller cities, though their access to the housing market in terms of home ownership has been far more mixed. A number of my key informants spoke of both individual success stories and the improvement of entire neighborhoods as a result of refugees moving in by purchasing homes and beautifying lawns and gardens and investing in renovations.

> Yeah, I mean, as I said, I think our recent refugees have actually brought better statistics to the high schools, graduation, college enrollment, college attendance, grade advancement... lots of small businesses, lots of participation in the workforce ... we've had a lot of home purchases by the Bhutanese populations, I mean, well over a hundred, I mean, it's in the hundreds. And they fix up these homes, they really fix up the whole streets and the whole

neighborhoods. I think it's good for the state culturally. I mean, people wouldn't say this but, not being from X, I think it's one of the ways that X is able to kind of join the global community.

- City Manager, Northeast

We have quite a few ethnic restaurants, we have other stores, grocery stores, and we have some other businesses that are … they cater primarily to their ethnic community but in other situations they can cater to a more mainstream population. So, we have that business situation. We have a lot of home ownership among refugees. We have increased the tax base, economically. I'm thinking, speaking specifically in economics here. The contributions that refugees make to the growth, to the economic vitality of the community is very measurable. It's a big part of how we make the neighborhood that they were placed in which was kind of falling apart, come back to life. We really expand our thinking as a community a great deal by having refugees present. The arts are more vital, cultural events are occurring weekly, international types of things that are happening in many cases because we have a diverse refugee population.

- State Refugee Coordinator, Northwest

A lot of the traditional gateway cities and older, bigger, cities are either experiencing loss of quality housing stock and availability, or it's becoming more expensive – I don't want to say "gentrification" – but knocking down old buildings and rebuilding housing that's more expensive; prohibitively so for refugees in the early days of resettlement. So, I think that's the biggest hardship for refugees in a big city. But I think this is becoming a problem in some of the smaller ones too. Refugees come in and fix up a place but can they actually stay there? Do they buy houses in the same neighborhood or do they get pushed out?

- State Health Coordinator, Northwest

This last quote raises an important question regarding the position of refugees in "improving" a given location. As pointed out previously, the research on refugee integration in Canada over the past decade highlights housing as a particular barrier to integration and success (Carter & Osborne, 2009; Carter & Vitiello, 2011; Hiebert & Mendez, 2008), with a majority of refugees unable to move from the rental sector to home ownership even after a decade in the country. There is little evidence from my study as yet that home ownership has become widespread amongst recently arrived refugees, though the perception based on highly publicized anecdotal evidence for some new arrivals (particularly the Bhutanese) remains high.

This perception of refugees as drivers of urban renewal is thus perhaps more important than the reality of such changes. In many of the locations that I have studied (whether in smaller or larger sites), refugees have tended to be resettled in those areas that are most likely to have less expensive housing that can accommodate large or extended families. This means that refugees are often put into direct competition for housing with a number of others (especially low-income groups and student populations). Several key informants referenced increasing tensions between newcomers and existing populations based on struggles to secure housing and access to services.

> A negative impact would be [the] impact on housing, perhaps that is already non-existent in a local community. Especially communities where people have "moved off the farm" into the community, so you have an older population occupying housing. Now you have young refugee families moving in and that kind of can be challenging for schools or other social services, or even trying to find housing for locals. I've seen a lot of resentment because of that.

- State Health Coordinator, Northwest

> Housing is the biggest issue in the resettlement world right now because it's just very limited in how much money we get to pay for their rent and all of their needs in the first three months. But the problem is people think refugees are getting all these free apartments and it's just not true. And especially if we rent with a landlord that's built a new building in a neighbourhood, even if it's just for a few months, then other people complain – people waiting for like Section 8 vouchers and things like that – that they're changing the neighbourhood to give homes to foreigners but there's nothing for them.

- Resettlement Agency Staff, Northeast

Additionally, several respondents noted a new dynamic in which refugees who had been earlier blamed by some low-income residents for being displaced from their neighborhoods becoming in turn pushed out by rising rents as the areas they had "improved" were in turn made unaffordable to them.

> One of the big problems I've noticed is with X Avenue. It's right in-between a neighbourhood that people would describe as gentrifying and another that is pretty middle-class. It's kind of the border between them. When the refugees first started coming, X Avenue was pretty run down. But they moved into it and they really fixed it up. But they didn't buy the houses, even if they were falling down, they couldn't afford to. But now it's a nice street – even nicer than the neighbourhood everyone is saying is getting gentrified so the landlords can jack up the rents and a lot of those refugees are getting priced out and they're pretty bitter about it.

- City Official, Northeast

### 5.3. *Refugees as a lifeline for declining cities*

Intimately connected to both of these previous themes is the third major narrative underlying the perceptions of refugee resettlement in smaller communities amongst the key respondents in our study, that of refugees as a lifeline for declining cities. While all three of these themes highlight a sense of refugees as adding a new or different element to the makeup of smaller cities, nowhere is the sense of need *for* refugees as apparent in the language of key respondents. One gets the sense that in the minds of these actors within the USRAP if smaller communities did not have refugees they simply would not survive. The discourse is at times desperate and reflects real demographic change and anxiety – the populations of such cities and towns are aging and in this view are thus unable to sustain economic and social vitality; refugees represent a lifeline to continued viability (Lee et al., 2015; Smith, 2012).

> Well this is really interesting right now with a lot of smaller cities dwindling in size as their own children move to the big city, or industry that has kept a smaller community alive for decades decides to close and move. So sometimes refugees actually keep communities alive, or keep businesses alive. So that is definitely one impact, supporting smaller communities, and reinvigorating local business. Refugees then would tend to open a little restaurant or community store that caters to their needs. So, you start to see business development in smaller communities.

- Resettlement Official, Federal Agency

When we first moved here, the only places who would rent to us were in the neighborhoods no one wanted to live in. They were falling apart – not full of crime but they were not good. We have made them better. Now everyone wants us in their neighborhoods and to rent their houses and even the city has a program to help our families buy houses because they know we will make the city better.

- Ethnic Community Based Organization (Refugee) official, Southeast

I don't care where we get people from. I'll take 'em from Syria, from California, from Africa, from Spain, from Mexico, from New York City, I don't care. We're losing people, all the young people and the families, they're all moving out and I'm afraid in a few more years there'll be more empty houses and boarded up businesses if we can't get these refugees to come. I'll take anyone, anyone who'll make my city look lived in again.

- City Mayor, Northeast

Sentiments like this were common amongst many respondents, that absent refugee arrivals, the future looks bleak for small cities in the US, especially those outside of the orbit of larger metropolitan areas. And yet the resulting demographic change heralds another and different set of anxieties amongst other parts of the population – that the newcomers and their potentially dissonant culture represents an unwelcome change and the decline of other familiar ways of life:

> I've had to deal with clients (not refugees) tell us that the city isn't theirs anymore. They ask us why we bring refugees and believe conspiracy theories about us wanting to change X into Baghdad.

- Social Service Provider, Midwest

> When we tried to pass a resolution to become a welcoming community, we encountered a lot of resistance from community activists who said that we were in cahoots with the big developers to push out low-income residents and that we were going to use refugees to live in the new apartment complexes so that if they protested we could just call them racists.

- City Official, Southeast

> It's not just housing. We've had a real problem with some new refugees getting jumped on payday or being shaken down for money, especially by other minorities. They see them as

competition for entry level jobs, housing, social services and think they get help that they can't.

- Resettlement worker, Northeast

This kind of backlash has real and long-term effects. The racist, anti-immigrant, xenophobic and Islamophobic rhetoric that is on the rise nation-wide can couple with other resentments regarding the lack of affordability and social inequality that challenge low-income and marginalized populations more generally. This can lead to greater levels of precarity and danger, as recent plots to commit violence against refugees and their supporters reveal (Lind, 2018; Park, 2018).

## 6. Conclusion

What then can we say overall about the current context of refugee resettlement in smaller cities and towns across the US? It is clear that have been placed in large numbers in newer destinations in recent decades, and yet the outcome of these patterns has been mixed. For every Utica, New York, Dayton, Ohio, or Bowling Green, Kentucky where refugees have been lauded by civic officials and resettlement workers for helping to open new businesses, purchasing and improving homes, and strengthening the local labor pool, there is a Lewiston, Maine, Erie, Pennsylvania, or Rutland, Vermont in which the anxieties felt by the local population led to backlash against new arrivals.

A concern for many scholars and policymakers regarding the placement of refugees in non-traditional metropolitan locations is the sense that these places do not have the capacity or resources to provide the kinds of supports necessary for integration. These might include language training, workforce development, sensitivity to cultural differences, the distinctions within as well as between different refugee groups, and access to housing, transportation, educational and other supports that are often crucial to the first weeks, months and even years of settlement into a new community. Such concerns have been a part of the discussion within the US refugee regime, amongst resettlement agencies and federal, state and local authorities (as well as refugee advocates) even before the current controversies.

Another issue that I have not addressed in this paper but should of course also be of primary concern – what has been the experience for refugees themselves entering into these new spaces of settlement? How have *they* felt as the embodiment of either hopes or fears? What do they think of being in a smaller rather than a larger city, farther and possibly more alienated from their friends, families and communities? Do they stay at all or do they move sooner or later to a more traditional immigrant enclave in a more recognizable metropolitan area? What is the reception they are receiving and have they been able to make productive, welcoming homes and new lives for themselves? These are questions that I take up in other aspects of this research but are also actively being explored by numerous other scholars (Abdo & Craven, 2018; Hagstrom et al., 2019; Nagel & Ehrkamp, 2017).

While the public discourse regarding refugee resettlement in the US has focused primarily on expressions of altruism and humanitarian embrace on the one hand and on xenophobia, fear, and anxieties regarding social and economic burdens on the other, the three sets of overlapping narratives regarding refugees discussed in this paper suggest that resettlement for many of these locations is as much about the needs of the places they arrive in as they do the needs of the refugees who are ending up there. Taken together these three themes suggest the idea of the refugee as the "good" and "productive" immigrant, one that can fit into traditional narratives of hard work, education, and urban renewal that have long underwritten the trajectories of upward social mobility for some (though certainly not all) immigrant populations in the US (Portes & Rumbaut, 2014; Roediger, 2005). It is also not surprising that so many of the respondents who play key roles in different facets of the USRAP might express such sentiments, not only because they may have a deeply held belief in the mission of refugee resettlement but also in response to rising xenophobia, anti-immigrant and anti-refugee policies at the federal level and general misrepresentations of the program by media, the culture industries and local politicians.

Yet it is worth asking what the impact of such apparently positive representations of refugees and their impacts on smaller communities might be. Prashad (2000) questions, for example, the ways that supposedly positive stereotypes of "model minorities" in Asian communities play into anti-black racism in the US; so too we might ask whether the tropes of "refugees-as-saviors" for smaller communities posit the refugee as inherently more deserving or legitimate than undocumented immigrants. And what is the impact upon refugees – psychologically, economically, culturally and politically – to be put perhaps literally into the line of fire as the symbol and embodiment of not only demographic change in a region, but of material change within a city? Such questions must remain at the heart of our understanding of the role of refugees in the changing landscapes of the US.

## Notes

1. Most recently this includes those who worked with US agencies overseas in Iraq and Afghanistan and subsequently found themselves under threat (US Department of State, 2017).
2. Such as those displaced by the 2010 earthquake in Haiti whose protected status has been renewed since 2014 on a semiannual basis (USCIS, 2017).
3. These would include the tens of thousands of Central American children who began arriving at the southern US border by early 2013 fleeing violence in their home countries (Rosenblum & Ball, 2016).

## Acknowledgments

The author would like to thank Birgit Glorius and Rene Kreichauf as well as the anonymous reviewers for their valuable feedback throughout the preparation of this article.

## Disclosure statement

No potential conflict of interest was reported by the author.

## Funding

This work was supported by the National Science Foundation [1359895].

## References

Abdelkerim, Abdelkerim, & Grace, Marty. (2012). Challenges to employment in newly emerging African communities in Australia: A review of the literature. *Australian Social Work*, *65*(1), 104–119. https://doi.org/10.1080/0312407X.2011.616958

Abdo, Diya, & Craven, Krista. (2018). Every campus a refuge: A small college's engagement with refugee resettlement.". *Migration and Society*, *1*(1), 135–146. https://doi.org/10.3167/arms.2018.010112

ACS. (2013). American Community Survey 2009–2013 five year estimates. https://www.census.gov/programs-surveys/acs/technical-documentation/table-and-geography-changes/2013/5-year.html

Aldana, Raquel. (2018). The travel ban in numbers: Why families and refugees lose big. *The Conversation*. Retrieved from http://theconversation.com/the-travel-ban-in-numbers-why-families-and-refugees-lose-big-99064

Annett, Evan. (2017, February 23). Trump's original immigration ban: How it was introduced and how the courts shut it down. *Globe and Mail*. http://www.theglobeandmail.com/news/world/us-politics/the-donald-trump-immigration-ban/article33822491/

Barneche, Kelly., & Joe. (2014). Faith and responses to displacement. *Forced Migration Review*, *48* (November 2015), 9. https://www.fmreview.org/faith/barneche-joe

Beier, Raffael, & Fritzsche, Jasmin. (2017). Refugees and the city: UN-habitat's new urban agenda. *Forced Migration Review*, *55*(2017), 29–30. https://www.fmreview.org/shelter/beier-fritzsche

Brandt, Jessica. (2017, June 29). Order from chaos: What the partially-reinstated travel ban means for refugees. *Brookings*. Thursday. https://www.brookings.edu/blog/order-from-chaos/2017/06/29/what-the-partially-reinstated-travel-ban-means-for-refugees/

Carter, Tom, & Osborne, John. (2009). Housing and neighborhood challenges of refugee re-settlement in declining inner city neighborhoods: A Winnipeg case study. *Journal of Immigrant and Refugee Studies*, *7*(3), 308–327. https://doi.org/10.1080/15562940903150097

Carter, Tom, & Vitiello, Domenic. (2011). Immigrants, refugees and housing. In Carlos Teixeira, Wei Lei, & Audrey Kobayashi (Eds.), *Immigrant geographies of North American cities* (pp. 91–111). Oxford University Press.

Chavez, Leo. (2013). *The Latino threat* (2nd ed.). Stanford University Press.

Clevenger, Casey, Derr, Amelia Seraphia, Cadge, Wendy, & Curran, Sara. (2014). How do social service providers view recent immigrants? Perspectives from Portland, Maine and Olympia, Washington. *Journal of Immigrant and Refugee Studies*, *12*(1), 67–86. https://doi.org/10.1080/15562948.2013.823533

Coughlin, Reed, Stam, Kathryn, & Kingston, Lindsey N. (2016). Struggling to start over: Human rights challenges for Somali Bantu refugees in the United States. *The International Journal of Human Rights*, *20*(1), 123–137. https://doi.org/10.1080/13642987.2015.1061237

Dahinden, Janine. (2016). A plea for the 'de-migranticization'of research on migration and integration. *Ethnic and Racial Studies*, *39*(13), 2207–2225. https://doi.org/10.1080/01419870.2015.1124129

Danso, Ransford. (2002). From 'there'to 'here': An investigation of the initial settlement experiences of Ethiopian and Somali refugees in Toronto.". *GeoJournal*, *56*(1), 3–14. https://doi.org/10.1023/A:1021748701134

Dauvergne, Catherine. (2008). *Making people illegal: What globalization means for migration and law*. Cambridge University Press.

Eastmond, Marita. (2011). Egalitarian ambitions, constructions of difference: The paradoxes of refugee integration in Sweden. *Journal of Ethnic & Migration Studies*, *37*(2), 277–295. https://doi.org/10.1080/1369183X.2010.521323

Feng, Hou. (2006). Spatial assimilation of racial minorities in Canada's immigrant gateway cities. *Urban Studies*, *43*(7), 1191–1213. https://doi.org/10.1080/00420980600711993

Gow, Greg. (2005). Rubbing shoulders in the global city: Refugees, citizenship and multicultural alliances in Fairfield, Sydney. *Ethnicities*, *5*(3), 386–405. https://doi.org/10.1177/1468796805054962

Hagstrom, Paul, Pereira, Javier, & Wu, Stephen. (2019). The happiness of refugees in the United States: Evidence from Utica, NY. *Journal of Refugee Studies*, fez026. https://doi.org/10.1093/jrs/fez026

Haines, Douglas, & Rosenblum, Karen. (2010). Perfectly American: Constructing the refugee experience. *Journal of Ethnic & Migration Studies*, *36*(3), 391–406. https://doi.org/10.1080/13691830903358445

Hardwick, Susan. (2011). The geography of whiteness: Russian and Ukrainian 'coalitions of color' in the Pacific Northwest. In John Frazier, Eugene Tettey-Fio, & Norah Henry (Eds.), *Race, ethnicity, and place in a changing America* (pp. 415–424). SUNY Press.

Hiebert, Daniel, & Mendez, Pablo. (2008). *Settling in: Newcomers in the Canadian housing market, 20012005*. CMHC.

Jacobsen, Karen, & Landau, Loren B. (2003). The dual imperative in refugee research: Some methodological and ethical considerations in social science research on forced migration. *Disasters*, *27*(3), 185–206. https://doi.org/10.1111/1467-7717.00228

Jones, Richard (Ed.). (2008). *Immigrants outside megalopolis: Ethnic transformation in the heartland*. Lexington Books.

Kallio, Kirsi P., Häkli, Jouni, & Pascucci, Elisa. (2019). Refugeeness as political subjectivity: Experiencing the humanitarian border. *Environment and Planning C: Politics and Space*.

Karas, Tania, & Campbell, Monica. (2020). *20 immigration issues to watch in 2020*. Public Radio International. https://www.pri.org/stories/2020-01-03/10-us-immigration-issues-watch-2020

Keck, Nina. (2016). *Vermont town debates Syrian refugee resettlement program*. Vermont Public Radio. https://www.npr.org/2016/07/01/484381701/vermont-town-debates-syrian-refugee-resettlement-program

Kennedy, Merrit. (2016, September 30). Texas pulls out of federal refugee resettlement program. *NPR*. http://www.npr.org/sections/thetwo-way/2016/09/30/496098507/texas-pulls-out-of-federal-refugee-resettlement-program

Kraly, Ellen, & Vogelaar, Peter. (2011). Starting with spoons: Refugee migration and resettlement programs in Utica, New York. In John Frazier, Eugene Tettey-Fio, & Norah Henry (Eds.), *Race, ethnicity, and place in a changing America* (pp. 389–400). SUNY Press.

Lee, Sungkyu, Choi, Sunha, Proulx, Laurel, & Cornwell, Jennifer. (2015). Community integration of Burmese refugees in the United States. *Asian American Journal of Psychology*, *6*(4), 333–341. https://doi.org/10.1037/aap0000027

Lind, Dara. (2018). *The conspiracy theory that led to the Pittsburgh synagogue shooting, explained: Xenophobic panic over the migrant caravan found a target – Jews who helped settle refugees*. https://www.vox.com/2018/10/29/18037580/pittsburgh-shooter-anti-semitism-racist-jewish-caravan

Loescher, Gil, Betts, Alexander, & Milner, James. (2008). *The United Nations high commissioner for refugees (UNHCR): The politics and practice of refugee protection into the twenty-first century*. Routledge.

Mackenzie, Catriona, McDowell, Christopher, & Pittaway, Eileen. (2007). Beyond 'do no harm': The challenge of constructing ethical relationships in refugee research. *Journal of Refugee Studies*, *20*(2), 299–319. https://doi.org/10.1093/jrs/fem008

Massey, Douglas. (2008). *New faces in new places: The changing geography of American immigration*. Russell Sage Foundation.

McBrien, J. Lynn. (2011). The importance of context: Vietnamese, Somali, and Iranian refugee mothers discuss their resettled lives and involvement in their children's schools. *Compare: A Journal of Comparative & International Education*, *41*(1), 75–90. https://doi.org/10.1080/03057925.2010.523168

Misra, Tanvi. (2019, January 31). The cities refugees saved. *CityLab*. https://www.citylab.com/equity/2019/01/refugee-admissions-resettlement-trump-immigration/580318/

Mitschke, Diane, Mitschke, Aaron, Slater, Holly, & Teboh, Consoler. (2011). Uncovering health and wellness needs of recently resettled Karen refugees from Burma. *Journal of Human Behavior in the Social Environment*, *21*(5), 490–501. https://doi.org/10.1080/10911359.2011.566466

Miyares, Ines. (1998). *The Hmong refugees experience in America: Crossing the river*. Routledge.

Morello, Carol. (2018, September 17). US slashes the number of refugees it will allow into the country. *Washington Post*. https://www.washingtonpost.com/world/national-security/us-slashes-the-number-of-refugees-it-will-allow-into-the-country/2018/09/17/a26f398a-bac1-11e8-bdc0-90f81cc58c5d_story.html?noredirect=on&utm_term=.39935c904cc5

Mott, Tamar. (2010). African refugee resettlement in the US: The role and significance of voluntary agencies. *Journal of Cultural Geography*, *27*(1), 1–31. https://doi.org/10.1080/08873631003593190

Nagel, Caroline, & Ehrkamp, Patricia. (2017). Immigration, Christian faith communities, and the practice of multiculturalism in the US South. *Ethnic and Racial Studies*, *40*(1), 190–208. https://doi.org/10.1080/01419870.2016.1229489

Nawyn, Stephanie. (2010). Institutional structures of opportunity in refugee resettlement: Gender, race/ ethnicity,and refugee NGOs. *Journal of Sociology and Social Welfare*, *37*(1), 149–167. https://scholarworks.wmich.edu/jssw/vol37/iss1/9

Neidell, Matthew, & Waldfogel, Jane. (2009). Program participation of immigrant children: Evidence from the local availability of head start. *Economics of Education Review*, *28*(6), 704–715. https://doi.org/10.1016/j.econedurev.2009.06.004

Nelson, Lise, Trautman, Laurie, & Nelson, Peter B. (2015). Latino immigrants and rural gentrification: Race,"illegality," and precarious labor regimes in the United States. *Annals of the Association of American Geographers*, *105*(4), 841–858. https://doi.org/10.1080/00045608.2015.1052338

Newberry, Jay, & Darden, Joe. (2011). Opposite sides of the road: An analysis of African-American and Somali-immigrant migration patterns in the U.S. In John Frazier, Eugene Tettey-Fio, & Norah Henry (Eds.), *Race, ethnicity, and place in a changing America* (pp. 108–128). SUNY Press.

Park, Madison. (2018). 3 men convicted in plot to attack Somali immigrants with weapons of mass destruction. *CNN*. https://www.cnn.com/2018/04/19/us/bomb-plot-kansas-conviction/index.html

Pho, Tuyet-Lan, Gerson, Jeffrey, & Cowan, Sylvia. (2007). *Southeast Asian refugees and immigrants in the mill city: Changing families, communities, institutions—thirty years afterward*. University Press of New England.

Portes, Alejandro, & Rumbaut, Ruben. (2014). *Immigrant America: A portrait*. University of California Press.

Prashad, Vijay. (2000). *Karma of brown folk*. University of Minnesota Press.

Rabrenovic, Gordana. (2007). When hate comes to town: Community response to violence against immigrants. *American Behavioral Scientist*, *52*(2), 349–360. https://doi.org/10.1177/0002764207306063

Roediger, David. (2005). *Working towards whiteness: How America's immigrants became white: The strange journey from Ellis Island to the suburbs*. Basic Books.

Rose, Damaris, & Ray, Brian. (2011). How gender matters to immigration and settlement in Canadian and US cities. In Carlos Teixeira, Wei Li, & Audrey Kobayashi (Eds.), *Immigrant geographies of North American cities* (pp. 138–157). Oxford University Press.

Rosenblum, Marc, & Ball, Isabel. (2016). Trends in unaccompanied child and family migration from Central America. *Migration Policy Institute*. Retrieved from http://www.migrationpolicy.org/research/trends-unaccompanied-child-and-family-migration-central-america

RPC. (2018). *Refugee admissions by region: Fiscal year 1975 through 31-Aug-2018*. Department of State, Office of Admissions – Refugee Processing Center. http://www.wrapsnet.org/admissions-and-arrivals/

Sanyal, Romola. (2012). Refugees and the city: An urban discussion. *Geography Compass*, *6*(11), 633–644. https://doi.org/10.1111/gec3.12010

Sherrell, Kathy, Hyndman, Jennifer, & Preniqi, Fisnik. (2005). Sharing the wealth, spreading the 'burden'? The settlement of Kosovar refugees in smaller British Columbia cities. *Canadian*

*Ethnic Studies, 37*(3), 76–96. https://link-gale-com.ezproxy.uvm.edu/apps/doc/A145680860/AONE?u=vol_b92b&sid=AONE&xid=f76647f4.

Singer, Audrey, Hardwick, Susan, & Brettell, Caroline. (2008). *Twenty-first-century gateways: Immigrant incorporation in suburban America*. Brookings Institution Press.

Singer, Audrey, & Wilson, Jill. (2011). From 'there to here': Refugee resettlement in metropolitan America. In John Frazier, Eugene Tettey-Fio, & Norah Henry (Eds.), *Race, ethnicity, and place in a changing America* (pp. 364–388)). State University of New York Press.

Smith, Yda J. (2012). Resettlement of Somali Bantu refugees in an era of economic globalization. *Journal of Refugee Studies, 26*(3), 477–494. https://doi.org/10.1093/jrs/fes039

Tang, Eric. (2015). *Unsettled: Cambodian refugees in the New York City hyperghetto*. Temple University Press.

Timberlake, Elizabeth, & Oahn Cook, Kim. (1984). Social work and the Vietnamese refugee. *Social Work, 29*(2), 108–113. https://doi.org/10.1093/sw/29.2.108

UNHCR. (2009). *UNHCR policy on refugee protection and solutions in urban areas*. https://www.unhcr.org/protection/hcdialogue%20/4ab356ab6/unhcr-policy-refugee-protection-solutions-urban-areas.html

UNHCR. (2019). *Resettlement in the United States*. http://www.unhcr.org/resettlement-in-the-united-states.html

US Department of State. (2017). *Special immigrant visas (SIVs) for Iraqi and Afghan translators/interpreters*. https://travel.state.gov/content/visas/en/immigrate/iraqi-afghan-translator.html

USCIS. (2017). *Temporary protected status designated country: Haiti*. https://www.uscis.gov/humanitarian/temporary-protected-status/temporary-protected-status-designated-country-haiti

Uteng, Tanu Priya. (2009). Gender, ethnicity and constrained mobility: Insights into the resultant social exclusion. *Environment and Planning A, 41*(5), 1055–1071. https://doi.org/10.1068/a40254

Yan, Yom. (2006). *New Americans, new promise: A guide to the refugee journey in America*. Fieldstone Alliance.

# Negotiating urban solidarities: multiple agencies and contested meanings in the making of solidarity cities

René Kreichauf and Margit Mayer

**ABSTRACT**

Many cities have adopted welcoming strategies, branding themselves as cities of welcome or of solidarity. Urban scholarship to date has interpreted these efforts either under the rubric of municipal governance reform or urban citizenship, frameworks which both sideline the role of civil society and social movements of refugees. Since these actors play crucial roles in negotiating the terms of solidarity, hospitality and inclusion, this paper brings together research perspectives from urban governance, civil society, and (migrant) mobilization literatures to gain a better understanding of the collaborative/competitive interactions between the key players engaged in this urban policy arena. This discussion reveals that the evolving practices and interrelations of municipalities, civil society actors and social movements of refugees imply opportunities, but also difficulties in building substantively welcoming arrival structures, highlighting the contested meaning of terms such as "solidarity city" in the contemporary constellation.

## Introduction

Today, in a context where most nation states in the global North/West have responded to rising numbers of people fleeing war, persecution or other threats to their lives with increasingly nationalist and exclusionary policies, cities have adopted explicit welcoming strategies, enhancing their integration programs, and providing urban solutions to the arrival, residency, inclusion and social protection of forced and other migrants[1] (Agustin & Jørgensen, 2019; Darling, 2017). Various labels and concepts such as Sanctuary Cities in the US, Cities of Refuge in the UK and, more recently, Solidarity Cities in continental Europe reflect the expansion and diversity of municipal policies, initiatives, and urban-based movements to accept, shield, and include (forced) migrants (Foerster, 2019). They share a desire to contest national immigration laws and policies and to foster solidarity by – often creatively – (re)imagining different ways of urban life and living together

(Baban & Rygiel, 2020). The idea and concept of solidarity is a central and inherent part of these policies and initiatives. However, it takes on different meanings, visions and practices for local policy making and the support of vulnerable migrants, depending on place-particular circumstances, national and urban contexts, political and administrative systems, actors, addressed populations, and policy goals (Bauder, 2017, 2020; Kron & Lebuhn, 2020). This raises the following questions: What privileges cities to facilitate the welcoming of migrants, and what specific material and political features allow for and result in urban "solidarity", and in what form? Who are the factions and actors that participate in and negotiate the development of the solidarity city, and what interests do they pursue?

Such questions have mostly been addressed by research coming either out of scholarship in urban governance, local immigration and asylum policies and regimes (Agustin & Jørgensen, 2019; Foerster, 2019; Scholten & Penninx, 2016; Werner et al., 2018), or under the rubric of urban (as opposed to national) citizenship (Bauböck, 2003; Bauder, 2017; Isin & Turner, 2007; Lebuhn, 2013). However, as the trope of the welcoming or "solidarity city" involves different types of agency and practical implications, these literatures are marked by blind spots and hamper a comprehensive and critical understanding of the sometimes cooperating, at other times conflicting actors in the making of solidarity cities (García, 2006; Jeffries & Ridgley, 2020). In particular, Bauder and Juffs (2020, pp. 46 f.) find that "in the migration and refugee literature, this multi-dimensional character of the concept of solidarity is not always acknowledged." Consequently, a differentiated analysis of solidarity (cities) must not only pay attention to the complexity and dynamics of solidarity concepts and practices – their various tensions, antagonisms, and contradictions –, but also to the multiplicity of actors (including local governments, welfare associations, churches, private entrepreneurs, nonprofits, volunteers, activists and forced migrants themselves) involved in the effort of making cities hospitable to newcomers.

This article brings together research perspectives from urban governance, civil society as well as migrant mobilization literatures to gain a fuller understanding of the collaborative/competitive interactions between the key players engaged in this urban policy arena. To do so, it applies Agustin's and Jørgensen's (2019) typology of three types of solidarity (institutional solidarity, civic solidarity, and autonomous solidarity) and conceptualizes solidarity city projects as "zones of negotiation" (Pott & Tsianos, 2018). With this approach, we analyze the encounters of and negotiations between the actors from the realms of the local state, that of civil society, and from the people who fled their homes. We apply Berlin as primary case (with some references to other German cities) to contextualize and explain how solidarity is practiced in its different forms as well as to disentangle the various actors and interests relevant for and engaged in the solidarity city. The combination of these three actors and types shows that their respective organizations, institutions, (infra)structures, motives and goals manifest rather different concepts of solidarity and social justice, which is an outcome of the ambiguous and contradictory ways in which solidarity unfolds, is organized, negotiated and practiced by different urban actors – ranging from neoliberal diversity management to self-organized horizontal participation and direct democracy practices. Thus, our paper provides a basis for evaluating the scope and multidimensional meaning of "a city of solidarity" in contemporary Europe. Thereby, it contributes to those recent literatures in urban studies that focus on the structures of migrant and refugee solidarity in European cities (Agustin & Jørgensen, 2019; Bauder, 2020; Dicker, 2017; Kron & Lebuhn, 2020; Siim & Meret, 2020) as well as to debates in (forced) migration studies on the local settlement of forced migrants in urban environments at large (Glorius & Doomernik, 2016; Kos et al., 2016; Muggah & Abdenur, 2018; Zapata-Barrero et al., 2017).

The article proceeds as follows. We first present our analytical framework in the context of a review of current scholarship in the field. After thus developing the conceptual approach for this study, we focus on three sets of actors relevant in the field – urban politicians and administrations, civic society actors such as the increasingly well-organized volunteers, and refugee activists – and their respective solidarity practices, highlighting their interactions and negotiations with each other. We conclude by discussing the role of these actors, their understandings of solidarity, and the challenges they face in making the solidarity city.

## Solidarity cities as zones of negotiation

The literature about (forced) migrant solidarity pays a considerable amount of attention "to look(ing) beyond or below the nation-state to understand how such solidarities are generated and how they may be able to transform and invent social norms, political subjectivities, and even institutional frameworks" (Agustin & Jørgensen, 2019, pp. 12 f.). The urban scale in particular receives dedicated attention as a space in which new forms of solidarity toward (forced) migrants emerge, are articulated and translated into particular practices (Baban & Rygiel, 2020; Jeffries & Ridgley, 2020). Here, we identify two major bodies of work that study solidarities regarding (forced) migrants in the urban context from different perspectives and research interests.

First, many urban scholars and political scientists "discuss the issue of solidarity in connection with citizenship and the making and unmaking of political communities" (Oosterlynck et al., 2016, p. 773). Using the idea of (urban) citizenship as a lens to understand urban solidarities, they study the efforts "cities" make to welcoming and including migrants with a focus on their ability to contest the exclusionary nature of national citizenship and to redefine citizenship in a more inclusive manner (Bauder, 2017; Isin & Turner, 2007; Lebuhn, 2013). In this body of work, the new policies of "solidarity cities" are interpreted as expressions of urban citizenship, which protect residents with precarious legal status and seek to expand their rights to access local services through policy innovations, new ordinances and regulations, and/or the creation of agencies (Isin & Nyers, 2014). In addition, a number of scholars in Europe concentrate on recent municipal governance reforms in the reception and integration policies of local authorities through urban governance perspectives including studies of local reception, accommodation and housing approaches (Adam et al., 2020; Kos et al., 2016; Lidén & Nyhlén, 2015; Neis et al., 2018; Seethaler-Wari, 2018; Steiner & Reinhard, 2018; Werner et al., 2018). These studies illustrate that urban solidarity and local accommodation/integration policies vary in different parts of the world and at the level of cities, because they are "embedded in rather different regional geopolitical situations, national legal regimes, local discourses of migration, and other place-particular circumstances" resulting in distinct political aims and diverse localized strategies (Bauder, 2020, p. 3). For example, sanctuary cities in the US focus mostly on illegalized migrants, whereas in the

UK and continental Europe policies pursue the goal of supporting forced migrants in the phase of arrival through local-level policies and practices (Bauder, 2017).

Sanctuary or solidarity city policies are "not the result of top-down policy making, but are put onto cities' agendas by social movement actors and through strong bottom-up mobilizations" (Kron & Lebuhn, 2020, p. 92). Thus, a second body of literature looks into bottom-up solidarity movements, the rise of forced migrant-led politics and their impact on the situation of (forced) migrants and on urban policy making (García, 2006). This includes scholarship in social movement and critical citizenship studies analyzing migrant solidarity activism, social movements, and political protest as part of a "new era of protest" (Ataç et al., 2015; Della Porta, 2018; Isin & Turner, 2007; Kron & Lebuhn, 2020; Rosenberger et al., 2018). To different extents, this work is concerned with "acts of citizenship" in order to discuss migrant solidarity as a political act, the way it leads to the emergence of political mobilizations, and how migrants as political actors have changed the notions of citizenship, solidarity and city-making (Isin & Nielsen, 2008; Nyers & Rygiel, 2012; Siim et al., 2018). This literature also includes civil society mobilizations, forms of civic engagement and pro-migrant solidarity practices as it studies "the processes of 'doing' solidarity and practicing resistance" (Siim & Meret, 2020, p. 2).

These top-down and bottom-up perspectives and discussions imply different concepts and understandings of solidarity in theory and practice, leading Agustin and Jørgensen (2019, p. 25) to argue that "solidarity is overloaded and in that sense is a 'floating signifier', i.e. a signifier that is open to continual contestation." Similarly, Bauder and Juffs (2020) find that solidarity escapes a single definition, as it depends not only on locally specific contexts and actors involved on the ground, but also on the perspectives and interests of those studying these processes. A number of scholars, however, seek to capture precisely this multidimensionality in the intricate contemporary engagements of solidarity. Oosterlynck et al. (2016, pp. 766 f.), for example, define four sources of solidarity, of which "encounter" and concrete interpersonal practices result in current innovative forms of solidarity, "which are located in relationally constituted places that become sites for everyday negotiation and agonism" (Oosterlynck et al., 2016, p. 777). Siim and Meret (2020) apply the concept of "reflective solidarity" for studying solidarity practices in Copenhagen and Berlin. Reflective solidarity conceives of solidarity "in terms of a process generated and continuously (re-)negotiated in a context of difference and multiple concrete particularities, rather than based on an ideal of pre-given agreement, shared views and expectations" (Siim & Meret, 2020, p. 3). Disagreement and differences are here seen as a basis of solidarity, mutual connection and inclusive dialogs on a collective level. Based on an intensive literature review, Bauder and Juffs (2020) use philosophical and conceptual underpinnings to develop a categorization scheme including several types of solidarity. The type "recognitive solidarity" is of particular interest to us, because it relates to practices of liberation (e.g. autonomist solidarity, volunteer and migrant activism, refugee protests) that challenge different forms of discrimination, domination and oppression and through which new political relations, subjectivities and progress emerge.

This overview of a select range of concepts currently applied in our field illustrates that solidarity may be understood as generative, inventive, relational and spatial (Agustin & Jørgensen, 2019; Kron & Lebuhn, 2020). Urban solidarity with forced migrants may take place on and connect different scales (from trans-local solidarities to city networks down to neighborhood and place-based activism), and it may generate relations between spaces and/

or actors (as, e.g. in alliance building). It can create political subjectivities and collective identities through the invention of new imaginaries and political relations and through spatial practices that connect different actors and transform or invent social norms and institutional structures (Siim & Meret, 2020). Most powerfully, solidarity may be a political project and an imagination, resulting from the engagements, intersections, tensions and negotiations of different actors involved (Kron & Lebuhn, 2020). Solidarity, thus, is the starting point *and* potentially the outcome of urban struggles, contestations, and negotiations among those actors.

To capture the interests and negotiations of diverse actors involved in solidarity city projects we turn to Agustin's and Jørgensen's (2019) actor-centered typology (institutional solidarity, civic solidarity, and autonomous solidarity) and conceptualize solidarity city projects as "zones of negotiations" (Pott & Tsianos, 2018). With this framework, we study the involvement, tensions and negotiations of local government, civil society, and forced migrant actors through which solidarity emerges – out of a "battleground" of colliding actors, interests and visions (Ambrosini, 2020). According to Agustin and Jørgensen (2019) institutional solidarity refers to the formalization of solidarity, the potential political action of solidarity and its regularization by institutions. It describes the capacity of enabling (infra)structures and policies to materialize solidarity maintaining the connections with civil society and (forced) migrant organizations. Autonomous solidarity includes autonomous and self-organized initiatives on the ground, often in the form of organized horizontal participation and direct democracy practices without a direct relationship to the local state. Civic solidarity encompasses civil society mobilization, forms of civic engagement and pro-migrant solidarity practices and activism – the "vast number of manifestations and actors, such as NGOs, local communities and individuals" that are "receptive to the idea that the vulnerabilities, which prevent people from participating on equal terms, must be eliminated" (Agustin & Jørgensen, 2019, p. 41). These solidarity types and actors encounter each other and, through constant efforts to bring about, shape and establish (different forms of) solidarity, they produce the solidarity city. This takes place because of continuous and manifold *solidarity practices* through which the three types of actors and their specific interests, perceptions, and strategies determine solidarity. Solidarity practices reveal asymmetric power relations; the contestation and negotiations of these relations result in different outcomes and material and political features of solidarity city projects. On the one hand, the "solidarity city" undergoes progressions; it is not static, but changes in accordance with constant negotiations and changing power and actor constellations. On the other hand, solidarity is in itself a transformative power that can change infrastructures, practices and institutions (Kreichauf et al., 2020; Pott et al., 2018; Pott & Tsianos, 2018). In what follows, we employ this typology in presenting the field of migrant support, thus elucidating the role played by each of the actors in co-creating the solidarity city.

## Negotiating the solidarity city

Obviously, municipal politicians and administrators and their respective departments and agencies play decisive roles in setting the parameters for a city's receptive climate. But in 2015, when thousands of forced migrants were suddenly present in cities such as Berlin, Munich or Hamburg without state agencies able to even register, let alone provide minimum standards of accommodating them, it became clear that municipalities are not the only actors shaping the welcoming climate. With established welfare organizations being equally overburdened, a non-traditional form of civic engagement in the form of volunteer activism arose – next to existing human rights, refugee, anti-racist, or urban movements – playing a crucial role in making cities sites of hospitality. Further, various organizations of migrant groups and their actions also contributed to the formation of hospitable environments. Out of civic engagement and migrant groups' activism a bottom-up Solidarity City movement has emerged in Berlin since 2015 with the aim to create safe and empowering conditions for forced migrants within their city, and to build support for the migrants' demands for the right to stay as well as the right to move.

In the following sections, we apply the relational and actor-centered approach described above in order to identify and unpack these different sets of actors (local governments, civil society actors and forced migrants) and their approaches to making solidarity cities. While we focus in this paper on the practices and negotiations of actors in Berlin, we also present findings and literatures from other cases to contextualize the Berlin developments and to reveal the core actors' abilities and practices for creating "solidarity cities".

### *Solidarity as policy: Municipal governments and their leeways*

To many, the city is a key site to counter the intensifying tendencies toward closure and for building an "open Europe." As primary locations and agents for forced migrant reception and integration (Doomernik & Ardon, 2018), European cities currently represent forms of institutional solidarity in two ways. They "scale up," articulate and formalize the generation of solidarity by building European-wide alliances in the form of city-networks, and they "scale down" the matter of migrant solidarity, inclusion and protection to particular local policy-making efforts at the urban scale by enabling (infra) structures and policies to materialize solidarity (often in cooperation with civil society actors) (Agustin & Jørgensen, 2019).

In Germany, the majority of cities present themselves as open, welcoming cities. Freiburg, for example, portrays itself as "city of refuge," while Hamburg highlights its cosmopolitan character by labeling itself "arrival city" (Schmidt, 2018). When the city of Oldenburg joined the EU-supported coalition of 10 European cities (URBACT-Program) called "Arrival Cities" (founded in 2015), its website referred to the goal of this network to "develop a community in which mutual tolerance and acceptance is being lived."[2] Other German cities have joined the network "Solidarity Cities," which was founded by the Eurocities Initiative in 2016. This alliance includes Barcelona, Naples, Athens, Thessaloniki, Amsterdam, Gdansk (i.e. initially port cities), and pressures EU institutions for a coordinated approach to what its founding document calls the "refugee crisis" (EUROCITIES, 2016). It sees cities as central for the reception and integration of forced migrants, and on that basis advocates for better – more effective as well as more humanitarian – reception of them across the EU. After a red-green-red coalition had come to power in Berlin in December 2016, its government also officially joined the Solidarity Cities network in 2019. More than 100 municipalities participate in the EU-sponsored *Intercultural Cities Programme* (ICC) in order to exchange and consult with each other about effective policy development in the area of refugee integration, Berlin's district Neukölln among them. And as a reaction to stranded migrants at the shores of

Italy, who had been rescued in the Mediterranean by civil rescue missions, twelve German cities founded an alliance "Cities of Safe Harbours" announcing their willingness to invite more than their regular quota.[3] Meanwhile, the mayors of 90 German as well as of other European cities such as Barcelona, Naples and Palermo have joined in declaring their willingness to take in more refugees directly, and their alliance continues to exert pressure on their national governments to support instead of blocking them (MiGAZIN, 2019, 2020).

Systematic evidence on the concrete functions played by such European (trans-) national city networks is still scarce. A comparative study of two Italian cities (Caponio, 2019) finds they play primarily symbolic functions, such as legitimizing local integration policy and positioning the city vis-à-vis other European cities and EU institutions. But locally, where politicians and administrations pledge – not least by officially joining such solidarity networks – to adopt a migrant-friendly policy vision, they often do enact protective and supportive measures, especially where they are held accountable by strong civil society actors insisting on making the "solidarity city" status real and meaningful (Kron & Lebuhn, 2020).

On the local level, cities have to navigate a "complex 'tightrope-walk', as they [city governments] share (and compete for) power with supralocal institutions, at the same time as having to respond to possible conflicts and demands emerging from below" (Bazurli, 2019, p. 344). As the "last instance of government", their proximity to problems of exclusion places them in a contradictory role regarding the granting of access to and take-up of rights (Gebhardt, 2016). This positioning of cities implies, on the one hand, the *potential* that municipalities may provide people with unclear residence status with access to rights and services by exerting their relative autonomy. Local authorities may do so because they are usually more concerned with maintaining public order in their jurisdictions than they are with the legal status of their residents; and they have particular "urban capabilities" to react autonomously to special conditions at the city level and to problems of everyday organization (Kuge, 2019). Therefore, they may be inclined to take more reasonable, pragmatic and favorable positions toward their undocumented residents (Kratz & Nowak, 2017). Barber (2013, p. 71) even claims that cities generally "prefer problem solving to ideology and party platforms," they have indeed frequently provided, in a kind of "logic of emergency," some, at least temporary, partial, or improvised form of membership based on human rights.

On the other hand, cities are tied into the implementation of federal politics of migration control (through registry offices, social services departments, schools, etc.) (Bazurli, 2019). Consequently, they have hardly any legal competence regarding the rights of forced migrants and their access to state institutions. However, the "local turn in migrant integration policies" – stipulated by the increased arrival of forced migrants since 2014 – has catapulted cities to the forefront of asylum politics introducing reception, accommodation and social policies that often go beyond or even circumvent the thicket of EU and national laws and guidelines regulating how municipalities shall implement asylum policies (Scholten & Penninx, 2016). Cities can exploit the contradictions, vagueness, and gaps in the laws governing the multi-scale system of asylum and immigration politics. They have broad leeway in how they interpret and implement the respective laws (Buckel, 2011). This leeway applies (as detailed elsewhere, see Mayer, 2017) not merely to areas where cities are able to work with greater flexibility and creativity than higher government units – and thus may, for example, assemble refugee taskforces in a matter of days, rapidly launch pilot projects for forced migrant reception, or coordinate different city services and external stakeholders. It applies also to the prescribed mandated tasks for which municipalities are constitutionally responsible within the federal division of labor, such as the issuing of residence titles.

So, even in areas, where higher levels of government prescribe in some detail the substance and procedure of the task, and where municipalities aren't supposed to have any margins of discretion, the frame set by national and state policies can be stretched. Some asylum and residency laws contain numerous ill-defined legal concepts, which "the state ministry either cannot or does not want to concretize with state law or direction," so that it remains up to local immigration offices to interpret federal or state stipulations (Schammann & Kühn, 2016, p. 11). Thus, for example, if an asylum request has been rejected, the staff of the municipal immigration agency in Germany, which is closely supervised by the Land government, can decide whether specific factors might prevent deportation of a rejected claimant, e.g. whether and for how long a temporary permit ("exceptional leave to remain") might be issued. Also, vague legal concepts allow German states (Länder) and cities (up to now) to refrain from withholding services when asylum seekers have "violated" their duty to cooperate with the asylum procedure, have not taken up suggested "work opportunities," or did not participate in an obligatory "integration class." Bendel (2016) finds that if the municipal leadership wants, it can prioritize integration over regulatory/policing imperatives.

Berlin has been utilizing this leeway in varied ways. Contradicting German federal law on mandatory housing of forced migrants in mass accommodation facilities (AsylVerfG, §51(1)), Berlin allows refugees to rent apartments after three months in a reception or accommodation center. Since 2011, public housing companies and the Senate have cooperated in providing such apartments exclusively for forced migrants; and Berlin's agency for refugee affairs (LAF) has installed a department that supports forced migrants in finding apartments in the regular housing market (Kreichauf, 2018). Further, the current red-green-red government's "Master Plan" for Integration and Participation of Refugees employs the term "refugee" also for people "whose application for asylum was rejected or who do not fulfill their enforceable obligation to leave the country," recognizing the variety of reasons that justify a right to stay (SenIAS, 2018, p. 12; cf. Endnote vi). In 2019, Berlin turned its foreigners' office (Ausländerbehörde) into a "welcome office" in the form of a state agency for immigration – the first of its kind in Germany –, heralding a "cultural change" regarding the reception and integration of migrants. However, at this point, it is unclear whether this rebranding actually has positive implications for migrants or serves merely symbolic politics (Amin, 2020).

Within the political science perspective of urban and local government and of public administration, the way municipalities make use of this latitude is primarily explained by factors such as the constellation of political parties in power, the financial wealth of the municipality, and the political will or orientation of its political leaders (De Graauw, 2016; Verhoeven & Duyvendak, 2017). The self-perception of a city is also recognized as playing a role for how creatively municipal autonomy is exerted: where open, cosmopolitan traditions are appreciated, even migrants with precarious legal status are granted access to public goods and services – be it through parallel structures via civil society organizations such as welfare associations, churches, and informal groups, or through incorporation into regular systems, where citizenship status or residence title are simply made irrelevant (Bendel, 2016; Buckel, 2011; Heuser, 2019).

Important as these factors are, they do not exhaust potential explanations for the differential uptake of municipal leverage. Because this perspective basically ignores the dynamic recent development of civic engagement, it misses out on an important explanatory variable. Within this perspective, civic engagement becomes visible and its influence measurable only through the lens of municipal actors: only if they experience political lobbying by local refugee support organizations as intense, is it viewed as possibly contributing to the municipalities exerting their leverage (Agustin & Jørgensen, 2019; Kron & Lebuhn, 2020). As a matter of fact, though, municipalities have over the last few years not only adapted and re-arranged their departments tasked with processing and integrating forced migrants but also developed new arrangements to direct and coordinate the civic engagement of volunteers, even where such civic groups were not intensely lobbying (e.g. Daphi, 2017; Gesemann et al., 2019; De Graauw, 2016). Against this backdrop, Verhoeven and Duyvendak (2017, p. 564) find that often "governmental players join[ing] forces with non-governmental players in contentious actions against policies they want to prevent or redress." Many German municipalities by now operate as steering, coordinating and guiding players within the asylum policy arena, often by outsourcing to non-state actors the implementation of large parts of the functions that – within the multi-scale division of labor for governing the refugee situation – fall on them. After the voluntary activists initially had pushed – spontaneously and from below – into a relative void of state provision, by now a panoply of new administrative functions, along with newly created (intermediating and coordinating) positions, and a range of new (funding) programs have evolved. This complex web of public, private and civil society actors and institutions, which now governs the arrival and inclusion of the new migrants, has been examined primarily in local case studies (e.g. Verhoeven & Duyvendak, 2017). Such research finds that the new state strategies for furthering and coordinating the civic engagement for forced migrants are often contested, particularly with regard to an appropriate balance between support and regulation of that engagement, as will be discussed in the next section.

Summing up, we can say that cities do utilize the instruments and options at their disposal to constructively deal with the challenges that arrive with large numbers of newcomers, including options that improve the inclusion of disenfranchised migrant groups. They possess a "unique combination of representing a level of governance that is local and thus able to represent pragmatism, efficiency and legitimacy, but at the same time being able to learn from each other through horizontal networking" (Doomernik & Ardon, 2018, p. 92). However, the different ways in which cities make use of their relative autonomy produce rather uneven infrastructures of local service provision in cities across Germany. Also, the presence and strength of two civil society actors plays at least as large a role in shaping the welcoming character of a city as the municipality-immanent factors highlighted by urban policy research in determining whether and to what degree cities merit such titles.

### *Solidarity as movement: civic engagement for forced migrants*

Civic forms of solidarity refer to the ways civil society initiatives work to include forced migrants in the host society. They are practiced by a vast number of actors (from well-structured NGOs to informal volunteers) that are not part of the local state, but that interact and often cooperate with it to push forward claims and strategies of solidarity. Civic solidarity thus forges "new alliances and collective identities in different kinds of spaces" – ranging from critical opposition toward government to efforts to influence policy-making (Agustin & Jørgensen, 2019, p. 41).

While Germany has a long tradition of migrant support groups, organizations and initiatives, their number has exploded in the summer of 2015, when hundreds of thousands of forced migrants arrived and soon faced cities overwhelmed by registering and taking care of them. To fill in, more than 150 initiatives sprang up in Berlin alone, civil society initiatives *next to* established welfare associations, nonprofit service providers, voluntary agencies, NGOs, refugee councils, and church communities, all of whose volunteer numbers jumped up as well (Daphi, 2016; Karakayali, 2018; Karakayali & Kleist, 2016). Even though over the last five years, this engagement has diminished somewhat, it remains relatively stable in increasingly differentiated forms, in a variety of fields, and on a higher level than in other spheres of social engagement (Gesemann et al., 2019). Its forms range from individual, personal "adoptions" of one or several refugees, to guide and assist them in navigating the unfamiliar and often challenging territory, all the way to collective efforts, in some cases even city-wide networks, providing and mobilizing for solidarity. For example, a German writer couple started an intercultural housing project for forced migrants, the so-called Refugio, in cooperation with the Berlin City Mission, a welfare association providing support for homeless people, and since 2015, also operating asylum shelters in Berlin. The five-story building provides living and working spaces for 40 people with and without forced migrant background, thus allowing newcomers to connect with long-established residents and providing opportunities for their social and economic integration (Kreichauf et al., 2020). Another example would be the city-wide network "Netzwerk Berlin Hilft", which describes itself as Berlin's most important "helpdesk," "information platform," and "thinktank." It works to coordinate and bring together forced migrants, volunteers, and administrative and governmental actors in the field.

While these grassroots initiatives differed significantly in their organizational culture from traditional volunteering in welfare associations and church groups, they have been and continue to be characterized by enormous heterogeneity, dispositions ranging from charity-based volunteerism to left-radical ambitions of *No border!* activism. At the same time, in this form of "recognitive solidarity" (Bauder & Juffs, 2020), the boundaries between what is often described as "apolitical," humanitarian volunteerism (which frequently sees itself as doing the legwork for local government's integration measures) and self-organized projects of state-critical activists (who connect their practical support for refugees with political demands for more rights for migrants, especially those with precarious legal status) are often fuzzy, shifting, or contested. Various studies find that even in the absence of ideological framing, politicization processes do take place (Daphi, 2017; Sinatti, 2019). Though politicized, these groups usually do not prioritize demands for open borders, but primarily seek to improve conditions of reception on the ground.

Such demands will, however, become more radical when refugees, who have been looked after by the initiatives and who have successfully "integrated," are to be deported: then demands easily become extended to include the right to global free mobility and the right to stay.

The primary addressee of these initiatives is the municipality, as its agencies are more accessible than other levels of government, and the threshold for participation in and negotiation with local governments is lower. On many issues, the groups seek to collaborate with municipal political and administrative actors, but often they take on a watchdog function, reminding the municipal actors of promises made, thematizing deficits in the treatment of refugees or in the administrative structures, and urging remediation. Through their activities and the infrastructures they built, the initiatives have often also spawned a mobilization of local civil societies against nationalist and racist groups. Almost every district of Berlin has at least one such citizens' initiative for the support of forced migrants within their area, often using the district's name as a label as in "Moabit Hilft" or "Kreuzberg Hilft." They offer direct help for forced migrants (language courses, escort to government offices, doctors or apartment visits, etc.), and they hold the authorities accountable for their actions (or the lack thereof) by constantly revealing the circumstances under which forced migrants live in Berlin.

As some of these initiatives provide services to which forced migrants are legally entitled (language classes, counseling, accommodation), their practice has created the same type of dilemma that has been encountered since the beginning of the withdrawal of the state from social reproduction in the 1980s. While the self-help and voluntary labor that pitches in where governmental agencies were supposed to deliver social services may indeed mitigate in emergencies, it comes with a series of problems in that necessary qualifications are lacking, the services cannot be provided in the necessary comprehensive, all-embracing manner, or governments use the unsalaried civic engagement as a pretext to reduce public services instead of adjusting the latter to the growing demand (Van Dyk & Misbach, 2016). Also measured by the intentions of the engaged themselves, their voluntary "helping out" sometimes has problematic implications: While their unbureaucratic aid may be better suited to meet the real needs and hardships of forced migrants, the really-existing disparities in power and resources have consequences. In an environment where social and political rights depend on citizenship status, volunteers (with German citizenship) and refugees cannot function as co-equals, no matter how much the volunteers want to realize equality. (Alas, not all volunteers strive toward equality: some helpers are over-keen, and refugees occasionally encounter lack of respect or get disempowered while being "cared for"). Particularly in conditions of mass accommodation, avoiding paternalism on the side of the volunteers often remains a theoretical desire, even where they attempt to operate on a level playing field (Cappiali, 2017; Jungk, 2016).

Under these difficult conditions, a variety of forms of cooperation have evolved in the course of the last few years between initiatives and local administrations. Their co-production of refugee services has tended to *consolidate* and professionalize the initiatives' projects. *Consolidation* means that the initially informal groups now have more clearly defined responsibilities and goals, in many instances they founded (registered) associations, created (paid) positions, or tied themselves to umbrella organizations, so as to facilitate receiving and managing donations and grants. Some have signed explicit cooperation agreements with local authorities and taken on the delivery of specific public tasks as mandatory, thus turning themselves into new agencies of nonprofit work. At the same time, *precarization* occurs because the new programs' expectations and requirements are oriented to the capabilities and resources of the established welfare associations and service providers, with whom the initiatives are now competing. Even though a variety of (primarily state-provided) funding opportunities have meanwhile become available for initiatives, the financial resources accessible to them are often inflexible and do not fit their concrete needs. For lack of sufficient funding and support, the initiatives thus often have to develop their own programs under uncertain and laborious conditions and cannot plan with any long-term perspective (Gesemann et al., 2019).

On top of all this, they often feel pressed to assert themselves "against senseless administrative action" ("Offener Brief," 2015, p. 1; cf. also Berliner Initiativen, 2018). Even before the tightening of the immigration law with the so-called "Migration Packet" of June 2019,[4] initiatives observed an increasingly restrictive interpretation of provisions in the law as well as illicit administrative action (from faulty rulings by the Federal Office for Migration and Refugees to erroneous Jobcenter decisions), forcing them to spend much of their energy on making sure forced migrants get treated correctly according to the law. Because of such experiences, some initiatives have intensified their function as "control agency" of the state, and they have developed a more pronounced political self-image: they publish critical statements, participate in Open Letters, and have even organized strike actions to protest against the instrumentalization of their voluntary work.[5] Others have become practical: In some cities, groups have begun to organize "citizen asylum" to protect forced migrants threatened with deportation. They announce publicly that they are about to hide such refugees from the authorities and call for fellow Germans to sign such declarations of civil disobedience. At the same time, they build, with their activities around concrete protective spaces, practical structures of solidarity.[6] They see their campaign for "citizen asylum" not merely as defending the migrants' right to stay, but also as part of the struggle for a "solidarity city".[7] Similarly, protests against deportations have frequently developed into practical solidarity, as for example, in the work of the *No Lager* groups and in coalitions against deportations. Case studies of such alliances in various cities reveal the productive opportunities for collaboration between engaged activists with secure residence status and migrants with precarious legal status (e.g. Hinger et al., 2018, on Osnabrück).

"Solidarity City Berlin" was formed in the fall of 2015 by different groups such as MediBüro Berlin (a group of medical professionals and volunteers seeking to provide health care to people without insurance since 1996), the local "citizen asylum" group, and migrant initiatives. Together with welcome initiatives, refugee councils, migrant organizations, church groups, leftist organizations and academics it launched the nation-wide network "Solidarity City" in 2017, which carries forward the grassroots struggle for social rights and urban citizenship of the new arrivals, connecting similar initiatives across many small and large cities. This network's goals go far beyond those of the network of city governments with the similar name (Solidarity Cities), as it seeks to develop "a city for all," in which "everyone shall have the right to live and work", "no matter what 'legal' and financial status they have."[8] Their concept of solidarity wants to extend rights and resources to *all* groups excluded from access.

As health care is one of those resources from which growing numbers of people, not merely illegalized migrants, are excluded, Medibüro pursues to provide free services to uninsured people in need of medical attention and it campaigns for political solutions (Medibüro, 2016). The story of its struggles reveals how, through persistent activism, voluntary engagement may achieve some success in and through its negotiations with the local authorities, while also shining light on some of the hurdles and blockages in the way of turning hospitality discourse into a reality: The initiative set up a network of about 130 medical professionals, who donate their labor and engage in fundraising for drugs, lab tests, surgeries, etc., thus performing about 1000 free medical treatments annually. But since such voluntary support remains insufficient to the needs, MediBüro proposed, already in 2005, an anonymous health card as a politically feasible way to establish health care access equivalent in scope to the services that forced migrants granted asylum receive (Neumann, 2019). For a long time, Berlin's Department of Health was not ready to support the health card model Medibüro suggested. However, the various stakeholders continued to lobby for the model, until, in 2018, the red-green-red coalition set aside €1.5 million annually to provide free health services to uninsured people, and established a Clearing Agency as a first stop for such patients (Kron & Lebuhn, 2020). Given the fund's financial cap, this achievement remains limited, and Medibüro has publicly chided the delayed implementation of the model, the limited number of included health providers and the limited funding. Its members organized a strike action in June 2019, where they rallied together with other solidarity groups and health organizations in front of Berlin's Health Department for the universal right to access health services (Medibüro, 2019). This case illustrates that even with a left-leaning government, and even on the city level, initiatives pushing for practical realizations of solidarity with precarious migrant groups face institutional and political obstacles in their effort to co-create comprehensive solutions together with city officials.

To sum up the findings of this section, the civic engagement that has powerfully contributed to making cities welcoming has also consolidated an indispensable role within the urban policy field of forced migrant reception and integration. In the process of building and intensifying collaborative programs and structures of service provision to vulnerable newcomers, a novel policy field has emerged where municipal actors seek to encourage and simultaneously steer and control the civic engagement. The interactions between municipal authorities and administrations on the one hand and the initiatives, even if increasingly professionalized and incorporated, on the other, frequently reveal divergent concepts of what solidarity with precarious migrant groups should look like and what forms it should take.

## *Autonomous solidarity: the self-organization of refugees*

Since the fall of 2012, European cities have seen an uptick in protest actions by (forced) migrants, which galvanized – with marches, demonstrations, encampments, hunger strikes and occupations – far more public attention than previously. Instead of taking on the role of needy victims and objects of aid organizations, they organized themselves in order to achieve access not only to human rights, but also social and political rights, as well as the right to free mobility (for an overview of these protests see Odugbesan & Schwiertz, 2018). Encompassing very heterogeneous people, these movements and protests represent both "recognitive" (Bauder & Juffs, 2020) and "reflective" forms of solidarity (Siim & Meret, 2020) as they often take place in autonomous and self-organized ways, making claims to solidarity, participation and rights, and contesting their (ascribed) outsider positions within society and the forms of domination and oppression related to this position (McNevin, 2006). Their initiatives are often run on the basis of direct democracy and through assemblies (Agustin & Jørgensen, 2019). They are premised on the concept of solidarity, which "in each case appears to be based [...] on the explicit premise that the initiatives [...] are open to anyone in need" (Dicker, 2017, p. 78).

Through their acts of protest as well as claim and space making, politicized groups of forced migrants constitute themselves as political subjects and act as citizens, "even when the law does not recognize them as such" (Nyers, 2010, p. 142). Cities are "a crucial site for the mechanisms through which movements form, disband, transform, or fail to form in the first place" (Nicholls & Uitermark, 2016, p. 227; also see Dicker, 2017). Forced migrants use the urban space "as a 'frontstage' for their struggles over citizenship" (Swerts, 2017, p. 379), even though adopting this form of visibility enhances a variety of risks.

One of the most prominent mobilizations of refugees in Germany was triggered by the suicide of the Iranian Mohammad Rahsepars in Würzburg in 2012, and the way the state dealt with it. In response, refugees set up a tent camp in the center of the city, demanding free mobility, an end to deportations, to humiliating treatment, and to being housed in desolate container camps. After several demonstrations and a hunger strike by Iranian refugees in Würzburg, about 50 refugees together with supporters began a 600 km march to Berlin on September 8, in direct violation of the law forbidding them to leave the district they have been assigned to. A month later, they erected a tent camp at Berlin's *Oranienplatz*, where about 80 refugees from Sudan, Macedonia, Iran and Afghanistan held out, with others living in a nearby squatted former school building. About 6,000 people joined their demonstration on October 13 to the German parliament. With the protest camp at Oranienplatz, the refugees succeeded to create both a communal space and a publicly audible voice. It also allowed them to forge social networks, alliances, and innumerable personal connections between refugees, supporters and residents (Ataç et al., 2015; Steinhilper & Ataç, 2019; Wilcke & Lambert, 2015).

In April 2014, the camp was cleared on the basis of an agreement negotiated with the city which was supposedly supported by all participants. It included the taking down of the tents at Oranienplatz and for the refugees to leave the squatted school building. But not all refugee groups had agreed to voluntarily leave, because the so-called "agreement" did not include housing options for the refugees cleared from the campsite, the redistribution of their asylum cases to Berlin, nor residence permits on humanitarian grounds, even though politicians had promised in the previous negotiations to grant these demands (Flüchtlingsrat Berlin, 2015). Thus, a few dozen refugees remained in the former school building with the intention of converting it into an international Refugee Center – already they were running language classes and arts projects, where new and old residents as well as locals could meet and cooperate. But in the face of a 10-day siege by massive police forces, a majority of the refugees left the school building. As these were gradually evicted from substitute accommodations that had initially been provided, and illegally exposed to homelessness (as the Social Court would later affirm[9]), the remaining 24 occupants of the school building lost all faith in the authorities. As meanwhile many new arrivals were claiming public attention, and about 100 newly arrived Syrian refugees were housed in (the evicted) part of the school building (now run as emergency shelter by the welfare association Johanniter, which in 2016 signed an operator contract),[10] the district government saw an opportunity to play out "new", less battle-tried refugees against those that had become bothersome to the authorities.[11]

Still, refugees in Berlin and all over Germany continued to alert the public to the degrading conditions in mass housing facilities and to the humiliating effects of the respective asylum laws – with protest camps in central squares, go-ins and occupations, social media campaigns and PR work, exhibits and conferences, and finally also hunger strikes (Ataç, 2016; Steinhilper & Ataç, 2019). With organized bus tours to reception centers, they reached out to new arrivals, informing them of the rights they are entitled to as well as about the protests, with the goal of recruiting more activists. Again and again, they converged from peripheral (rural, ex- and suburban) locations of mass reception centers onto urban central spaces, churches, and vacant buildings, which not only empowered and visualized the protesting refugees but also allowed access to resources and networks that helped in their mobilization. Their cause became publicly present, especially where they employed spatial strategies such as occupations and claiming of spaces that transformed specific – public as well as enclosed – locations into social and political spaces.

But the authorities' response to refugee movements has been divisive and demoralizing in its effects. Especially where the movements have been strong, the strategy adopted by the authorities was to offer conditional concessions, which resulted in splitting and weakening the movement. Those who don't pick up on the concessions can more easily be evicted and criminalized. Due to such tactics of attrition and division and because of the increasingly more restrictive laws governing irregular migrants, the spectacular forms of activism that unfolded at the height of German "welcome culture" have become rarer. However, even under most difficult conditions more and (mostly) less visible forms of self-organization of refugees continue to take place. For example, the residents of an intake center outside Osnabrück organized themselves in response to the state of Lower Saxony's intensifying deportations: "with its own committee, its nightly patrol, and an alarm system of whistles" (Hinger & Kirchhoff, 2019; Maestro, 2017). The barracks housed about 300 Sudanese migrants, whose shared language and previous political organizing experience which made the mobilization of the blockades easier. When the police marched up in the night, the residents assembled in front of the former barracks, blocked the entrances with garbage cans and barricaded the gates with bicycles – and sang. Besides such actions designed to prevent the identification of the sought-after deportee/s, they also organized demonstrations in front of city hall to make their demands known: "Uninterrupted sleep! Better food! Reduced public transport fares!" Local authorities broke up this self-organization by changing their accommodation practices, eliminating the notice of imminent deportations, and arranging for a transfer of the Sudanese activists, thus deliberately undermining the organizing efforts of the refugees (Hinger & Kirchhoff, 2019; Hinger et al., 2018).

The particular mix of regulatory/repressive and social/humanitarian-oriented responses that have evolved in any particular city differs, of course, and as such has been shaping the opportunity structures for the struggles of irregular migrants in different ways (Ambrosini, 2020). Even though the concerns raised by the refugees are generally the same as those raised by welcome initiatives and other NGOs active in this field, even the same as those formulated by the pertinent agencies in municipalities that brand themselves as "welcoming cities," their deprivation of rights and thus their structural exclusion hampers equivalent recognition. But it is striking that almost everywhere strategies of "Divide and Rule!" have been employed: a few, usually short term and/or symbolic concessions are made to selected groups of protesting refugees, which is then used to ignore, repress, or criminalize the rest. In Berlin, even after a red-green-red government drafted an explicitly progressive plan for the integration and participation of refugees, the voices and demands of the refugees and many of their advocates continue to be ignored in important aspects, from ending the exclusion of those with limited residence permits from the entitlement to social housing, and stopping the ongoing deportations to countries such as Iraq, to the demand for uniform quality standards to be implemented across Berlin's different refugee accommodations (Berlin, 2018, 2019; Berliner Initiativen, 2018; Memarnia, 2017, 2018).

The successes won despite such difficulties are likely attributable to the dense networks of self-organization of the refugees, and to their embeddedness in sustainable supporter networks in the respective cities. Strong cross-movement linkages have formed in the course of the last few years in joint struggles against austerity cuts and the affordable housing crisis. Such settings are conducive for a convergence of diverse urban struggles and for forming inclusive solidarity practices, as they have frequently materialized in right to the city actions and joint campaigns of tenants and migrants (Hamann & Vollmer, 2019).

## The city as a place of solidarity?

This article explained the abilities and practices of three of the contributors to implementing forms of institutional, civic and autonomous solidarity relevant for making the city a site of welcome and protection. Our analysis shows that the constant negotiations within and between these actors allow solidarity to emerge, and that exclusionary state and municipal practices are intensely contested. By studying these practices and their interactions, we found that municipal institutions indeed play a vital role for the emergence of the "solidarity city". Municipalities often use their own resources and leeway for developing pragmatic approaches, sometimes guided by humanitarian concerns. Most often they enroll diverse local partners and civil society actors in these efforts and attempt to steer and coordinate these to varying degrees (Verhoeven & Duyvendak, 2017). However, it is civic engagement, grassroots and refugee movements – their persistent activism and frequently critical publicity – that push for solidarity and bring these issues to the municipal agenda and beyond. They are indispensable in creating the solidarity city (Kuge, 2019).

Our analysis of the struggles for and negotiations over solidarity also revealed the ambiguous and contradictory ways in which solidarity unfolds, is mediated, practiced and contested by these three distinct urban actors. Comparing the goals and practices of grassroots and movement groups to those of alliances and networks such as Arrival

Cities, Solidarity Cities or the *Intercultural Cities Programme* mentioned above (in the section on "Solidarity as policy"), important differences stand out. Civic and autonomous solidarity movements demand social justice, the right to free mobility and the right to be present/sedentary – regardless of any formal (urban) citizenship or residence status (Squire & Darling, 2013). Affiliated civic movements such as Solidarity City forcefully fight for equal rights, inclusion, urban citizenship, and for a "solidarity for all" (including the extension of rights and resources for all excluded and disfranchised groups). The Solidarity Cities network, on the other hand, describes itself on its website as an "initiative for managing the refugee crisis" and one of its primary activities is to lobby the European Commission to increase funding for social infrastructures particularly in those cities, where most of the migrants arrive and live.[12] Similarly, cities participating in the ICC Program benefit from the Council of Europe's support, which provides experts for advice and exchange on "emerging issues" such as human rights, anti-discrimination, diversity, refugee integration, and urban citizenship, as well as possibilities for cities to learn from each other in these fields.[13] Though the ICC Program deplores neoliberal austerity policies abstractly, it is silent on the slashing of allowances to refugees during their asylum procedure to below current welfare levels, pointing instead to the voluntary civic engagement as if that could make up for welfare provisions by the state. Thus, while the municipal governments joining these networks do advocate for migrant-friendly policies across the EU, they primarily seek to develop – and to garner EU funding for – locally effective solutions for "the refugee crisis." Their primary interest is in regulating the presence of irregular migrants, securing their social rights is not their major concern.

Even where cities do strive toward expanding migrants' rights through introducing novel local instruments that extend social participation rights to all residents, achieving social protection or social justice remains – under current conditions – unrealistic. This is because the material resources required for any substantive participation are usually not there. Implementing local ID programs as some U.S. cities have endeavored requires enormous expenditures, and even more is required to provide the material dimensions of participation, i.e. access to education and training, to the health system, to mobility and to affordable housing (Lebuhn, 2016, 2018). Given these conditions, it should not be surprising that even the most well-intentioned "welcome cities" explore and apply various differentiating categories for different groups more and less "worthy" of their hospitality. Besides the realities of austerity, there are also political realities, which Doomernik and Ardon point to, "cities can act in exclusionary ways too if electoral realities force them to" (Doomernik & Ardon, 2018, p. 93). With anti-immigrant parties and racist movements growing – also in large cities such as Berlin – "local policies of exclusion" may increasingly aim to exclude migrants and to separate them from the access to local social policy resources (Ambrosini, 2020).

For forced migrants, claiming rights and solidarity presupposes visibility, but this increases the risk of criminal pursuit, especially for those who do not (yet) have some form of residence status. Where refugees have appeared in public with collective actions or otherwise made their presence in the city visible, they were sometimes able to successfully employ this visibility as political strategy. But the local authorities' counter-strategies of policing and controlling, and using negotiations to divide, individualize and demoralize the activists, and also the seemingly innocuous bureaucratic strategy of inventing more and more categories to differentiate them, have the effect that visibility can become a great risk and an existential threat (Zetter, 2007). Even more so, since in the current political climate, where right-wing populism has become widespread, individual visibility intensifies the danger of becoming targets of racist aggression.

Not least for these reasons, the support and solidarity from welcome groups equipped with citizenship status is so crucial. But, as shown above, civic solidarity is not only made up of very heterogeneous groups, but also unevenly distributed. On top, their incorporation into the field of governing refugee arrival and integration has affected these groups and organizations in ways that make solidaristic support on eye level more difficult. The professionalization and monetization of volunteer refugee work that came hand in hand with its public recognition and appreciation have, in many cases, divested this work of its solidarity character. At the same time, many of these groups have experienced the incremental tightening of asylum politics and deportation practices as *lack* of recognition of their work and have reacted either with protest or retreat. To this day, the inclusion/incorporation of civic engagement in the provision of urgently needed services is celebrated by many sides as innovative solution. And this in spite of indications that the new state programs designed to strengthen civic engagement and to encourage and expand volunteer labor in refugee support, may be merely the latest variant of what Rose (1996, p. 328) has called "*governing ... through ... community.*" That means that a form of (austerity) governing that uses the "refugee crisis" as pretext for installing unpaid or underpaid local volunteer labor instead of trained, professional social workers, translators, nurses, doctors, and other skilled personnel – with detrimental effects for refugees: Their status as subjects with legal entitlements erodes as they morph into recipients of charitable aid and become dependent on the arbitrary cycles of people's willingness to donate and their readiness to engage in civic and community action.

In sum then, looking at the agencies of three of the central actors in the newly evolved field of urban refugee politics revealed how, in their interactions and negotiations, the differences in power relations matter as much as larger constraints and forces, such as austerity pressures or the privatization and restructuring of welfare states. These constellations explain the rise of solidarity cities in reaction to urgent municipal problems and as a movement, as well as the limits to the potential of solidarity cities. In Berlin, where under a progressive government in 2019 1,003 deportations were performed and 5,767 refugees left "voluntarily" (SenInnSport Senatsverwaltung für Inneres und Sport Berlin, 2020), some have argued that the litmus test for efforts to making Berlin a Solidarity City is its role in opposing deportations (Neumann, 2019, p. 29). Maybe the movements have not exerted enough pressure on local government to stop deportations, but the developments presented here make clear that without engaging in supra-local, national and EU level struggles, the realization of migrants' rights can only go so far, the power of cities remains circumscribed. Against this backdrop, European cities and urban movements might learn from sanctuary city policies and movements in the U.S., where growing numbers of cities seek to protect their undocumented and vulnerable residents by Don't Ask, Don't Tell (DADT) policies, refuse to cooperate with deportation procedures, or extend their rights through municipal I.D. cards (even though, in practice, these policies are not always successful and beneficial for migrants) (Bauder, 2019; Lebuhn, 2018). One thing our analysis has shown is that, in order to make the most out of the potential power of cities, it would be helpful to acknowledge how concepts (and limits) of solidarity have themselves become a site of antagonistic political interests and practices.

## Notes

1. In the current contribution, we use the term "forced migrant" in the context of the manifold causes and reasons that lead people to migrate (thus, going beyond the legal definition of "refugee"), and as a general notion for people who – as a consequence of economic, environmental, political, social drivers, or of their decision to migrate – end up in asylum procedures. Where we use the term "refugee", this is less to denote the legal definition of a refugee, but to refer to a political identity and agenda linked to struggles and claims for space and rights.
2. Retrieved from https://www.oldenburg.de/startseite/leben-umwelt/soziales/zuwanderung-und-integration/migration-und-teilhabe/arrival-cities.html, translated by authors.
3. https://seebruecke.org/startseite/sichere-haefen-in-deutschland/kongress-sichere-haefen/
4. https://www.spiegel.de/politik/deutschland/geordnete-rueckkehr-gesetz-was-steckt-im-migrationspaket-a-1271323.html
5. See, for example, http://freundstattfremd.de/wp-content/uploads/2016/10/Nicht-in-unserem-Namen-Ehrenamtlichen-Streik-11-11-2016.pdf
6. https://aktionbuergerinnenasyl.de
7. https://solidarity-city.eu/de/2017/12/04/faq-haeufige-fragen-zur-initiative-buergerasyl/
8. https://solidarity-city.eu/en/about/
9. See the Decision of the Social Court of 28 August 2014: http://www.fluechtlingsinfo-berlin.de/fr/pdf/SG_Berlin_AsylbLG_gekuerzt_und_Nachweis_Unterkunft_Oplatz_Lampedusa.pdf
10. http://berlin-hilft.com/2016/08/gerhart-hauptmann-schule-wird-nun-doch-fluechtlingsunterkunft/
11. http://wer.oplatz.net/news-from-ohlauer-school-wer-13/
12. See https://solidaritycities.eu/
13. See www.coe.int/en/web/interculturalcities

## Disclosure statement

No potential conflict of interest was reported by the authors.

## ORCID

René Kreichauf http://orcid.org/0000-0003-0004-6749

## References

Adam, Francesca, Föbker, Stefanie, Imani, Daniela, Pfaffenbach, Carmella, Weiss, Günther, & Wiegandt, Claus-C. (2020). Municipal housing strategies for refugees. Insights from two case studies in Germany. In Birgit Glorius & J. Jeroen Doomernik (Eds.), *Geographies of asylum in Europe and the role of European localities* (pp. 201–224). Springer.

Agustin, Óscar García, & Jørgensen, Martin Bak. (2019). *Solidarity and the 'Refugee Crisis' in Europe*. Palgrave McMillan.

Ambrosini, Maurizio. (2020). The local governance of immigration and asylum: Policies of exclusion as a battleground. In Maurizio Ambrosini, Manlio Cinalli, & David Jacobson (Eds.), *Migration, borders and citizenship. migration, diasporas and citizenship* (pp. 195–215). Palgrave Macmillan.

Amin, Nina. (2020). *Namenswechsel in Berlin. Wenn aus der Ausländerbehörde ein Einwanderungsamt wird*. RBB24. https://www.rbb24.de/politik/beitrag/2020/01/umbenennung-auslaenderbehoerde-landesamt-einwanderung-berlin.html.

Ataç, Ilker. (2016). Refugee protest camp Vienna: Making citizens through locations of the protest movement. *Citizenship Studies*, *20*(5), 629–646. https://doi.org/10.1080/13621025.2016.1182676

Ataç, Ilker, Kron, Stefanie, Schilliger, Sarah, Schwiertz, Helge, & Stierl, Maurice (2015). Struggles of migration as in-/visible politics. Introduction. *Movements. Journal Für Kritische Migrations- Und Grenzregimeforschung*, *1* (2), 1–18. http://movements-journal.org/issues/02.kaempfe/01.ataç,kron,schilliger,schwiertz,stierl–einleitung~en.pdf

Ataç, Ilker, & Steinhilper, Elias (2016, September 17). *Escaping from asylum to act as citizens: Political mobilization of refugees in Europe*. Open Democracy/ISA RC-47: Open Movements. https://www.opendemocracy.net/en/escaping-from-asylum-to-act-as-citizens-political-mobilization-of-refuge/.

Baban, Feyzi, & Rygiel, Kim. (2020). Introduction: Living with others: Opening communities to newcomers. In Feyzi Baban & Kim Rygiel (Eds.), *Fostering pluralism through solidarity activism in Europe. Palgrave studies in altruism, morality, and social solidarity* (pp. 3–29). Palgrave Macmillan.

Barber, Benjamin. (2013). *If mayors ruled the world: Dysfunctional nations, rising cities*. Yale University Press.

Bauböck, Rainer. (2003). Reinventing urban citizenship. *Citizenship Studies*, *7*(2), 139–160. https://doi.org/10.1080/1362102032000065946

Bauder, Harald. (2017). Sanctuary cities: Policies and practices in international perspective. *International Migration*, *55*(2), 174–187. https://doi.org/10.1111/imig.12308

Bauder, Harald. (2019). Urban sanctuary in context. In Jonathan Darling & Harald Bauder (Eds.), *Sanctuary cities and urban struggles. Rescaling migration, citizenship and rights* (pp. 25–49). Manchester UP.

Bauder, Harald (2020). Urban solidarity: Perspectives of migration and refugee accommodation and inclusion. *Critical Sociology*. E-print ahead of publication, June 2020. https://doi.org/10.1177/0896920520936332.

Bauder, Harald, & Juffs, Lorelle. (2020). 'Solidarity' in the migration and refugee literature: Analysis of a concept. *Journal of Ethnic and Migration Studies*, *46*(1), 46–65. https://doi.org/10.1080/1369183X.2019.1627862

Bazurli, Raffaele. (2019). Local governments and social movements in the 'refugee crisis': Milan and Barcelona as 'cities of welcome'. *South European Society and Politics*, *24*(3), 343–370. https://doi.org/10.1080/13608746.2019.1637598

Bendel, Petra. (2016). *Was Flüchtlinge brauchen – Ein Win-Win-Projekt. Ergebnisse aus einer Befragung in Erlangen*. FAU University Press.

Berliner Initiativen. (2018). *Stellungnahme der Initiativen zum Gesamtkonzept Integration und Partizipation Geflüchteter*. http://berlin-hilft.com/2018/12/17/stellungnahme-der-initiativen-zum-gesamtkonzept-integration-und-partizipation-gefluechteter/

Buckel, Simone. (2011). Urban Governance und irreguläre Migration: Städtische Politik als Handlungsraum im Konfliktfeld irreguläre Migration. In Oliver Frey & Florian Koch (Eds.), *Die Zukunft der europäischen Stadt* (pp. 246–262). Verlag für Sozialwissenschaften.

Caponio, Tiziana.(2019). *If mayors ruled migration: Promises and gaps*. Migration Policy Centre blog. https://blogs.eui.eu/migrationpolicycentre/mayors-ruled-migration-promises-gaps/.

Cappiali, Teresa M. (2017). 'Whoever decides for you without you, s/he is against you!': Immigrant activism and the role of the Left in political racialization'. *Ethnic and Racial Studies*, *40*(6), 969–987. https://doi.org/10.1080/01419870.2016.1229487

Daphi, Priska. (2016). Zivilgesellschaftliches Engagement für Flüchtlinge und lokale 'Willkommenskultur'. *Aus Politik und Zeitgeschichte (APUZ) 14–15*, 35–39. https://www.bpb.de/apuz/223923/engagement-fuer-fluechtlinge

Daphi, Priska. (2017). Zur Kooperation zwischen Behörden und Zivilgesellschaft in der Unterstützung Geflüchteter. *Forschungsjournal Soziale Bewegungen. Analysen zu Demokratie und Zivilgesellschaft*, *30*(3), 34–45. https://doi.org/10.1515/fjsb-2017-0058

Darling, Jonathan. (2017). Forced migration and the city Irregularity, informality, and the politics of presence. *Progress in Human Geography*, *41*(2), 178–198. https://doi.org/10.1177/0309132516629004

De Graauw, Els. (2016). *Making immigrant rights real: Non-profits and the politics of integration in San Francisco*. Cornell University.

Della Porta, Donatella (ed.). (2018). *Solidarity mobilizations in the 'refugee crisis'*. Palgrave Macmillan.

Dicker, Sophie. (2017). Solidarity in the city: Platforms for refugee self-support in Thessaloniki. In Juliano Fiori & Andrea Rigon (Eds.), *Making lives: Refugee self-reliance and humanitarian action in cities* (pp. 73–103). Humanitarian Affairs Team, Save the Children.

Doomernik, Jeroen, & Ardon, Djoeke. (2018). The city as an agent of refugee integration. *Urban Planning*, *3*(4), 91–100. https://doi.org/10.17645/up.v3i4.1646

EUROCITIES. (2016). *Refugee reception and integration in cities*. Eurocities. http://www.eurocities.eu/eurocities/events/Refugee-reception-and-integration-in-cities-WSPO-A8KCQ5. https://nws.eurocities.eu/MediaShell/media/RefugeeReport_final.pdf

Flüchtlingsrat, Berlin e.V. (2015). *Flüchtlingsproteste Oranienplatz und Gerhart-Hauptmann-Schule – Chronologie, Forderungen, Dokumente*. http://www.fluechtlingsinfo-berlin.de/fr/pdf/Chronologie_Oranienplatz.html

Flüchtlingsrat, Berlin e.V. (2018, December 17). *Gesamtkonzept zur Integration und Partizipation von Geflüchteten in Berlin: Viele Worte statt Taten*. https://fluechtlingsrat-berlin.de/wp-content/uploads/pm_gesamtkonzept_17dez2018.pdf

Flüchtlingsrat, Berlin e.V. (2019, October 19). *Flüchtlinge in Wohnungsnot. Berlin schließt zahlreiche wohnungslose Geflüchtete weiterhin vom Wohnberechtigungsschein aus*. https://fluechtlingsrat-berlin.de/presseerklaerung/06-10-2019-fluechtlinge-in-wohnungsnot-berlin-schliesst-zahlreiche-wohnungslose-gefluechtete-weiterhin-vom-wohnberechtigungsschein-aus/

Foerster, Amy. (2019). Solidarity or sanctuary? A global strategy for migrant rights. *Humanity & Society*, *43*(1), 19–42. https://doi.org/10.1177/0160597618817456

García, Marisol. (2006). Citizenship practices and urban governance in European cities. *Urban Studies*, *43*(4), 745–765. https://doi.org/10.1080/00420980600597491

Gebhardt, Dirk. (2016). Re-thinking urban citizenship from a policy perspective: The case of Barcelona. *Citizenship Studies*, *20*(6–7), 846–866. https://doi.org/10.1080/13621025.2016.1191431

Gesemann, Frank, Seidel, Alexander, & Mayer, Margit. (2019). *Entwicklung und Nachhaltigkeit von Willkommensinitiativen*. vhw (Bundesverband für Wohnen und Stadtentwicklung.

Glorius, Birgit, & Doomernik, Jeroen. (2016). Refugee reception in Europe: New insight into European localities. *Journal of Refugee Studies*, *29*(4), 429–439. https://doi.org/10.1093/jrs/few041

Hamann, Ulrike, & Vollmer, Lisa. (2019). Mieter*innenprotest in der postmigrantischen Stadt. *Forschungsjournal Soziale Bewegungen*, *32*(3), 364–378. https://doi.org/10.1515/fjsb-2019-0043

Heuser, Helene. (2019). *Kommunale Spielräume zur Förderung legaler Zufluchtswege. Bundeszentrale für politische Bildung*. Bundeszentrale für politische Bildung.

Hinger, Sophie, Kirchhoff, Maren, & Wiese, Ricarda. (2018). ‚We belong together!' Collective anti-deportation protests in Osnabrück. In Sieglinde Rosenberger, Nina Merhaut, & Verena Stern (Eds.), *Protest movements in asylum and deportation* (pp. 163–184). IMISCOE/Springer Open.

Hinger, Sophie, & Kirchhoff, Maren. (2019). Andauerndes Ringen um Teilhabe. Dynamiken kollektiver Proteste gegen Abschiebung in Osnabrück (2014-2017). *Forschungsjournal Soziale Bewegungen*, *32*(3), 350–363. https://doi.org/10.1515/fjsb-2019-0042

Isin, Engin F., & Nielsen, Greg M. (Eds.). (2008). *Acts of citizenship*. Zed Books.

Isin, Engin F., & Nyers, Peter (Eds.). (2014). *Routledge handbook of global citizenship studies*. Routledge.

Isin, Engin F., & Turner, Bryan S. (2007). Investigating citizenship: An agenda for citizenship studies. *Citizenship Studies, 11*(1), 5–17. https://doi.org/10.1080/13621020601099773

Jeffries, Fiona, & Ridgley, Jennifer. (2020). Building the sanctuary city from the ground up: Abolitionist solidarity and transformative reform. *Citizenship Studies, 24*(4), 548–567. https://doi.org/10.1080/13621025.2020.1755177

Jungk, Sabine. (2016). Willkommenskultur: Von neuen Chancen, alten Fehlern und Versäumnissen; Ein Essay. *Widersprüche: Zeitschrift für sozialistische Politik im Bildungs-, Gesundheits- und Sozialbereich, 36*(141), 99–108. https://nbn-resolving.org/urn:nbn:de:0168-ssoar-63823-2

Karakayali, Serhat. (2018). *Ehrenamtliches Engagement für Geflüchtete in Deutschland.* Institut für Migrationsforschung und Interkulturelle Studien der Universität Osnabrück.

Karakayali, Serhat, & Kleist, Olaf (2016). *Strukturen und Motive der ehrenamtlichen Flüchtlingsarbeit in Deutschland. 2. Forschungsbericht: Ergebnisse einer explorativen Umfrage vom November/Dezember 2015.* Berliner Institut f. empirische Integrations- und Migrationsforschung.

Kos, Sanne, Maussen, Marcel, & Doomernik, Jeroen. (2016). Policies of exclusion and practices of inclusion: How municipal governments negotiate asylum policies in the Netherlands. *Territory, Politics, Governance, 4*(3), 354–374. https://doi.org/10.1080/21622671.2015.1024719

Kratz, Bruce, & Nowak, Jeremy. (2017). *The new localism. How cities can thrive in the age of populism.* Brookings.

Kreichauf, René. (2018). From forced migration to forced arrival: The campization of refugee accommodation in European cities. *Comparative Migration Studies, 6*(7), 1–17. https://doi.org/10.1186/s40878-017-0069-8

Kreichauf, René, Rosenberger, Olivia, & Strobel, Paul. (2020). The transformative power of urban arrival infrastructures: Berlin's refugio and Dong Xuan center. *Urban Planning, 5*(3), 44–54. https://doi.org/10.17645/up.v5i3.2897

Kron, Stefanie, & Lebuhn, Henrik. (2020). Building solidarity cities. From protest to policy. In Feyzi Baban & Kim Rygiel (Eds.), *Living with others. Fostering cultural pluralism through citizenship politics* (pp. 81–105). Routledge.

Kuge, Janika. (2019). Uncovering sanctuary cities. Between policy, practice, and politics. In Jonathan Darling & Harald Bauder (Eds.), *Sanctuary cities and urban struggles. Rescaling migration, citizenship, and rights* (pp. 50–76). Manchester University Press.

Lebuhn, Henrik. (2013). Local border practices and urban citizenship in Europe: Exploring urban borderlands. *CITY, 17*(1), 37–51. https://doi.org/10.1080/13604813.2012.734072

Lebuhn, Henrik (2016, December). *Bilanz des Kommunalen Personalausweises in New York City. Zeitschrift LuXemburg. Gesellschaftsanalyse und linke Praxis, 3*, 114–119. Vorstand der Rosa-Luxemburg-Stiftung. https://www.zeitschrift-luxemburg.de/kommunaler-perso-new-york-city/

Lebuhn, Henrik. (2018). Stadtbürgerschaft 'Light'. Migration und Vielfalt in der neoliberalen Stadt. *Prokla - Zeitschrift für kritische Sozialwissenschaft, 48*(2), 325–333. https://doi.org/10.32387/prokla.v48i191.90

Lidén, Gustav, & Nyhlén, Jon. (2015). Reception of refugees in Swedish municipalities: Evidences from comparative case studies. *Migration and Development, 4*(1), 55–71. https://doi.org/10.1080/21632324.2014.894178

Maestro, Andrea. (2017, June 28). *Der Alarmschläger. taz.* http://www.taz.de/!5421431/

Mayer, Margit. (2017). Cities as sites of refuge and resistance. *European Urban and Regional Studies, 25*(3), 232–249. https://doi.org/10.1177/0969776417729963

McNevin, A. (2006). Political belonging in a neoliberal era: The struggle of the sans-papiers. *Citizenship Studies, 10*(2), 135–151. https://doi.org/10.1080/13621020600633051

Medibüro (2016). *Zwanzig Jahre Medibüro.* https://medibuero.de/wp-content/uploads/sites/10/2016/03/20_Jahre_Medibuero_Berlin_WEB.pdf

Medibüro. (2019, June 20). *Gesundheit für alle! Anonymisierter Krankenschein jetzt! Für ein solidarisches Berlin! Pressemitteilung.* Medibüro. https://medibuero.de/pm_warnstreik_17062019/

Memarnia, Susanne (2017, December 10). *Eine Wende ist das noch nicht. Der rot-rot-grüne Senat lässt etwas weniger abschieben, dennoch gibt es Kritik vom Flüchtlingsrat. taz.* https://taz.de/Fluechtlingspolitik/!5464986/

Memarnia, Susanne. (2018, July 9). *Integration von Flüchtlingen in Berlin. Das Konzept ist im Transitlager. taz.* https://taz.de/Integration-von-Fluechtlingen-in-Berlin/!5516140/

MiGAZIN. (2019, June 19). *Kommunen gründen Bündnis "Städte Sicherer Häfen".* https://www.migazin.de/2019/06/17/kommunen-buendnis-staedte-sicherer-haefen/

MiGAZIN. (2020, January 21). *Uneinigkeit über Städte-Initiative zur Flüchtlingsaufnahme.* Ekrem Şenol. https://www.migazin.de/2020/01/21/koeln-integrationsminister-uneinigkeit-staedte-initiative/

Muggah, Robert, & Abdenur, Adriana E. (2018). *Refugees and the City. The Twenty-first-century Front Line. World Refugee Council Research Paper* No. 2. Centre for International Governance Innovation.

Neis, Hans J., Meier, Briana, & Furukawazono, Tomoki. (2018). Welcome city: Refugees in three German cities. *Urban Planning, 3*(4), 101–115. https://doi.org/10.17645/up.v3i4.1668

Neumann, Mario. (2019). A solidarity city in the making. Berlin's government and left movements are pushing for migrants' social rights. In Wenke Christoph & Stefanie Kron (Eds.), *Solidarity Cities in Europe* (pp. 19–36). Rosa Luxemburg Stiftung. https://www.rosalux.de/fileadmin/rls_uploads/pdfs/sonst_publikationen/Broschur_SolidarischeStaedte.pdf

Nicholls, Walter, & Uitermark, Justus. (2016). *Cities and social movements: Immigrant rights activism in the US, France, and the Netherlands, 1970-2015.* Wiley-Blackwell.

Nyers, Peter. (2010). No one Is Illegal between City and Nation. *Studies in Social Justice, 4*(2), 127–143. https://doi.org/10.26522/ssj.v4i2.998

Nyers, Peter, & Rygiel, Kim (Eds.). (2012). *Citizenship, migrant activism and the politics of movement.* Routledge.

Odugbesan, Abimbola, & Schwiertz, Helge. (2018). "We are here to stay!" Refugee struggles in Germany between unity and division. In Sieglinde Rosenberger, Nina Merhaut, & Verena Stern (Eds.), *Protest Movements in Asylum and Deportation* (pp. 185–203). IMISCOE/Springer Open.

*Offener Brief von flüchtlingspolitischen Initiativen/Willkommensinitiativen im Land Brandenburg (open letter by refugee support groups/welcome inititives in the state of Brandenburg)* (2015, December). https://www.fluechtlingsrat-brandenburg.de/wp-content/uploads/2015/12/14_12_2015_Offener_Brief_Initiativen.pdf

Oosterlynck, Stijn, Loopmans, Maartens, Schuermans, Nick, Vandenabeele, Joke, & Zemni, Sami. (2016). Putting Flesh to the Bone: Looking for solidarity in diversity, here and now. *Ethnic and Racial Studies, 39*(5), 764–782. https://doi.org/10.1080/01419870.2015.1080380

Pott, Andreas, & Tsianos, Vassilis S. (2018). Verhandlungszonen des Lokalen. Potentiale der Regimeperspektive für die Erforschung der städtischen Migrationsgesellschaft. In Jürgen Oßenbrügge & Anne Vogelpohl (Eds.), *Theorien in der Raum- und Stadtforschung – Einführungen* (pp. 116–135). Westfälisches Dampfboot.

Pott, Andreas, Rass, Christoph, & Wolff, Frank. (2018). *Was ist ein Migrationsregime? What is a migration regime?* Springer.

Rose, Nikolas. (1996). Death of the social? Refiguring the territory of government. *Economy & Society, 25*(3), 327–356. https://doi.org/10.1080/03085149600000018

Rosenberger, Sieglinde, Merhaut, Nina, & Stern, Verena (Eds.). (2018). *Protest movements in asylum and deportation.* IMISCOE/Springer Open.

Schammann, Hannes, & Kühn, Boris. (2016). *Kommunale Flüchtlingspolitik in Deutschland* (translation by the authors). Friedrich-Ebert-Stiftung.

Schmidt, Katharina. (2018). *"Arrival Cities" – Ankommen in Deutschland.* Berlin: Bundeszentrale für Politische Bildung. https://www.bpb.de/politik/innenpolitik/stadt-und-gesellschaft/216878/konzept-ankunftsstadt-

Scholten, Peter, & Penninx, Rinus. (2016). The multilevel governance of migration and integration. In Blanca Garcés-Mascarenas & Rinus Renninx (Eds.), *Integration processes and policies in Europe* (pp. 91–108). IMISCOE Research Series.

Seethaler-Wari, Shahd. (2018). Urban planning for the integration of refugees: The importance of local factors. *Urban Planning, 3*(4), 141–155. https://doi.org/10.17645/up.v3i4.1696

SenIAS (Senatsverwaltung für Integration, Arbeit und Soziales). (2018). *Comprehensive Programme for the Integration and Participation of Refugees.* https://www.berlin.de/lb/intmig/themen/fluechtlinge/fluechtlingspolitik/

SenInnSport Senatsverwaltung für Inneres und Sport Berlin. (2020). *Antwort auf die Schriftliche Anfrage Nr. 18/22058 vom 09. Januar 2020 über Asylzugänge und Abschiebungen im Jahr 2019.* https://pardok.parlament-berlin.de/starweb/adis/citat/VT/18/SchrAnfr/s18-22058.pdf

Siim, Birte, & Meret, Susi. (2020). Patterns of reflective solidarity and migrant resistance in Copenhagen and Berlin. *Critical Sociology.* E-print ahead of publication, July 2020. https://doi.org/10.1177/0896920520944517.

Siim, Birte, Saarinen, Aino, & Krasteva, Anna. (Eds.). (2018). *Citizens activism and solidarity movement in contemporary Europe: Contending with populism.* Palgrave Macmillan.

Sinatti, Giulia. (2019). Humanitarianism as politics: Civil support initiatives for migrants in Milan's Hub. *Social Inclusion, 7*(2), 139–148. https://doi.org/10.17645/si.v7i2.1968

Squire, Vicki, & Darling, Jonathan. (2013). The 'minor' politics of rightful presence: Justice and relationality in the city of sanctuary. *International Political Sociology, 7*(1), 59–74. https://doi.org/10.1111/ips.12009

Steiner, Hans, & Reinhard, Antje. (2018). A culture of welcome: Challenges and experiences in the city of Freiburg/Germany. In Helmut Kury & Sławomir Redo (Eds.), *Refugees and migrants in law and policy* (pp. 379–392). Springer.

Steinhilper, Elias, & Ataç, Ilker. (2019). Contentious subjects: Spatial and relational perspectives on refugee mobilizations in Europe. In Katharyne Mitchell, Reece Jones, & Jennifer Fluri (Eds.), *Handbook on critical geographies of migration* (pp. 345–357). Edward Elgar.

Swerts, Thomas. (2017). Creating space for citizenship: The liminal politics of undocumented activism. *International Journal of Urban and Regional Research, 41*(3), 379–395. https://doi.org/10.1111/1468-2427.12480

Van Dyk, Silke, & Misbach, Elène. (2016). Zur politischen Ökonomie des Helfens. Flüchtlingspolitik und Engagement im flexiblen Kapitalismus. *Prokla, 46*(2), 205–227. https://doi.org/10.32387/prokla.v46i183.109

Verhoeven, Imrat, & Duyvendak, Jan Willem. (2017). Understanding governmental activism. *Social Movement Studies, 16*(5), 564–577. https://doi.org/10.1080/14742837.2017.1338942

Werner, Franziska, Haase, Annegret, Renner, Nona, Rink, Dieter, Rottwinkel, Malena, & Schmidt, Anika. (2018). The local governance of arrival in Leipzig: Housing of asylum-seeking persons as a contested field. *Urban Planning, 3*(4), 116–127. https://doi.org/10.17645/up.v3i4.1708

Wilcke, Holger, & Lambert, Laura. (2015). Die Politik des O-Platzes. (Un-)Sichtbare Kämpfe einer Geflüchtetenbewegung. *Movements. Journal für kritische Migrations- und Grenzregimeforschung, 1*(2), 1–23. https://movements-journal.org/issues/02.kaempfe/06.wilcke,lambert-oplatz-k%C3%A4mpfe-gefl%C3%BCchtete-bewegung.html

Zapata-Barrero, Ricard, Caponio, Tiziana, & Scholten, Peter. (2017). Theorizing the 'local turn' in a multilevel governance framework of analysis: A case study in immigrant policy. *International Review of Administrative Sciences, 83*(2), 241–246. https://doi.org/10.1177/0020852316688426

Zetter, Roger. (2007). More labels, fewer refugees: Remaking the refugee label in an era of globalization. *Journal of Refugee Studies, 20*(2), 172–192. https://doi.org/10.1093/jrs/fem011

# Index

Note: Page numbers followed by "n" denote endnotes.